Dyes and Pigments - Insights and Applications

Edited by Brajesh Kumar

Published in London, United Kingdom

Dyes and Pigments - Insights and Applications
http://dx.doi.org/10.5772/intechopen.104138
Edited by Brajesh Kumar

Contributors
Sundaramurthy Devikala, Johnson Maryleedarani Abisharani, Jorge Bañuelos-Prieto, Alaitz Peñafiel, Ainhoa Oliden-Sánchez, Edurne Avellanal-Zaballa, Leire Gartzia-Rivero, Rebeca Sola-Llano, Shahid Adeel, Wafa Haddar, Mahwish Salman, Abdul Ghaffar, Mehwish Naseer, Muhammad Usama, Manel Ben Ticha, Chi-Wai Kan, Kelvin Sin-Wang Kwan, Ranulfo Combuca da Silva Junior, Katieli da Silva Souza Campanholi, Flávia Amanda Pedroso de Morais, Laura Adriane de Moraes Pinto, Fabiana dos Santos Rando, Magali Soares dos Santos Pozza, Wilker Caetano

First published in London, United Kingdom, 2023 by IntechOpen
IntechOpen is the global imprint of INTECHOPEN LIMITED, registered in England and Wales, registration number: 11086078, 5 Princes Gate Court, London, SW7 2QJ, United Kingdom

British Library Cataloguing-in-Publication Data
A catalogue record for this book is available from the British Library

Additional hard and PDF copies can be obtained from orders@intechopen.com

Dyes and Pigments - Insights and Applications
Edited by Brajesh Kumar
p. cm.
Print ISBN 978-1-83768-113-6
Online ISBN 978-1-83768-114-3
eBook (PDF) ISBN 978-1-83768-115-0

For EU product safety concerns:
IN TECH d.o.o., Prolaz Marije Krucifikse Kozulić 3, 51000 Rijeka, Croatia,
info@intechopen.com or visit our website at intechopen.com.

Meet the editor

Dr. Brajesh Kumar is an assistant professor and head of the Post Graduate Department of Chemistry, TATA College, Chaibasa, India. He received a Ph.D. in Chemistry from the University of Delhi, India. His research interest is the development of sustainable and eco-friendly techniques for nanoparticle synthesis and their applications for environmental remediation; active films of organic solar cells; nanomedicine; sensors; natural product extraction, purification, and analysis; natural polymers; peptide chemistry; microwave and ultrasound-assisted organic synthesis; and organic synthesis. Dr. Brajesh Kumar has been credited for different national and international fellowships and has worked as a faculty member at various universities in India, Ecuador, and South Korea. He has also published numerous research articles and is an active reviewer of more than fifty journals. He is also listed among the top 2% of scientists worldwide by Stanford University, USA.

Contents

Preface

Color, which contributes so much to the beauty of nature, is vital to the attraction and acceptability of most products used by modern society. As long ago as the 25th century BC, man colored his surroundings and clothes using a limited range of natural colorants of both animal (Cochineal) and vegetable (Indigo, Alizarin, Tyrian purple) origin. Dyes and pigments are substances that, once applied to a substrate, lead to selective reflection or the transmission of incident daylight. Most natural dyes are found in the roots, barks, leaves, bracts, flowers, skins, and shells of plants. They are categorized as organometallic and organic compounds and exhibit low solubility in organic solvents. As a result, they are essentially found in the solid state during processing and when applied to the substrate.

This book is a collection of research and review papers on the chemistry and physics of dyes, pigments, and their intermediates, including chemical constituents, spectroscopic aspects, surface, solution, crystal formation, photochemical, and ecological or biological properties. It is a useful resource for a wide variety of researchers worldwide whose work involves dyes and pigment synthesis, imaging, sensors, energy, medicine, polymers, food products, toxicological properties, and more.

Dr. Brajesh Kumar
TATA College,
Kolhan University,
Chaibasa, Jharkhand, India

Chapter 1

A Revisit of the Underlying Fundamentals in the Laser Emission from BODIPYs

Alaitz Peñafiel, Ainhoa Oliden-Sánchez,
Edurne Avellanal-Zaballa, Leire Gartzia-Rivero,
Rebeca Sola-Llano and Jorge Bañuelos-Prieto

Abstract

This chapter aims to provide a comprehensive assessment of the laser performance of commercially available laser dyes based on the boron-dipyrromethene (BODIPY) chromophore in a liquid state, as well as to remark the main underlying photophysical signatures triggering such photonic behavior. First, we describe their light absorption and fluorescence properties in solution. This spectroscopic study is supplemented with quantum mechanics calculations and electrochemical measurements. Afterward, the dyes are tested as active media of tunable lasers under transversal pumping. The recorded laser efficiencies and photostabilities are correlated with the registered photophysical properties identifying the main structural guidelines and photonic parameters, which rule the laser bands' position, intensity, and stability. As a result, we provide a comparative dataset of the laser performance, not available hitherto. Besides, the unraveling of the complex molecular structure-photophysics-laser relationship should help in the rational design of new tunable dye lasers with an improved photonic response along the entire visible region and reaching eventually the near infrared.

Keywords: organic dyes, BODIPY, photophysical properties, fluorescence, laser

1. Introduction

Fluorescent organic dyes have emerged as suitable and highly recommended photoactive media in many light-driven devices and photonic applications overall [1–11]. Their success relies mainly on their chemical versatility, thanks to the recent advances in organic synthesis [12, 13]. Among them, functional fluorophores are in the spotlight. These pivotal molecules feature chromophoric cores, whose photophysical signatures can be finely modulated by tailored chemical modifications of the molecular structures through well-established synthetic protocols. Therefore, a given molecular scaffold can be applied for different and selective purposes just adjusting the molecular structure by a rational design of the fluorophores. The search and generation of sophisticated fluorophores has boosted dye chemistry as a powerful tool for tunable photonics. Indeed, there are several families of organic dyes (such as coumarins, acridines, pyronines, rhodamines, oxazines, and styryl and cyanines-based dyes), with absorption/fluorescence bands along the whole visible spectral region, which could be suitable scaffolds to this aim [2]. However, there are still some problems that should be fixed. For instance, coumarins are endowed with low photostability, while styryl- and cyanine-based dyes usually display modest fluorescence efficiencies owing to their push–pull chromophores. Such photonic drawbacks are solved in other dyes, but their access is not straightforward from a synthetic point of view; their purification is tedious or the options for post-functionalization are limited. Therefore, the challenge in dye chemistry is to find out a bright and stable molecular platform easily accessed and amenable to a wide sort of synthetic routes to decorate the chromophore with functional groups, which enable a deep but controlled modulation of their photonic performance.

The archetype of modern fluorophores is the renowned BOron DIPYrromethene dyes (trademark BODIPY) [14–16]. This cyclic cyanine features a dipyrrin (or dipyrromethene) chromophoric core chelated by a difluoroboron bridge (**Figure 1**) [17], which offsets most of the aforementioned shortcomings from both a photonic and synthetic point of view. The first reports of BODIPY date from the sixties [18], but it was not until the early nineties when they gain recognition as photoactive media of tunable lasers, yielding high long-lasting efficiencies [19, 20]. After such pioneer work, BODIPY witnessed a booming growth and many researchers paid attention to this fluorophore and incorporate it as the cornerstone of their research projects [21–28]. As a result, nowadays BODIPY is qualified as a top-ranked dye successfully applied in biotechnology, and a myriad of optoelectronic and photovoltaic devices. Indeed, there is a huge chart of BODIPYs reported and patented, bearing exhaustive chemical transformations and functionalizations for specific photonic demands.

As aforementioned, BODIPYs became renowned owing to their laser performance. Since then, several attempts were conducted to modulate the laser emission wavelength [29] and ameliorate the efficiency and photostability [30, 31], including their confinement in solid hosts to develop solid-state dye lasers (SSDL) [32, 33]. Actually, some BODIPY laser dyes are available commercially. Exciton is one of the main

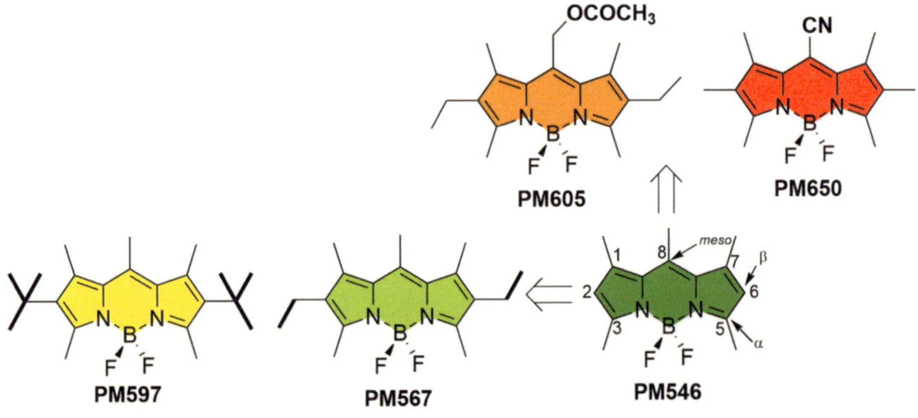

Figure 1.
Molecular structure of the BODIPY laser dyes purchased by exciton.

stockiest and provides technical sheets with information about their laser performance (maximum wavelength, tunability range, conversion efficiency, and stability as half lifetime) at different concentrations (up to millimolar), environments (liquid and solid), and pumping lasers. Indeed, the available BODIPYs from Exciton are named as PyrroMethenes followed by their laser emission wavelength (for instance, **PM546**, **PM567**, **PM597**, **PM605**, and **PM650** in **Figure 1**). However, in some cases, important data are missed (like photostability in **PM605** and **PM650** or efficiencies in some solvents for the rest of the dyes), and for each BODIPY, the data are collected under different experimental conditions or setups. Therefore, it is difficult to establish a well-grounded comparison between them to outline master guidelines about the molecular structure–laser relationship.

To offset such shortcoming and enable a direct comparison of the laser performance of these reference BODIPYs, we have measured their laser properties at the same experimental conditions, transversally pumped with a wavelength-tunable optical parametric oscillator (OPO) coupled to the third harmonic (355 nm) of the Nd:YAG laser. Furthermore, we aim to understand the complex interplay between molecular structure and photonic properties, a basic knowledge, which should enable the development of improved laser dyes. To accomplish this task, we have revisited their photophysical properties and conducted an exhaustive computationally aided spectroscopic and electrochemical study. This analysis is focused on identifying the main photophysical parameters, which triggers the laser signal, mainly the efficiency and photostability, as the key properties to determine the workability of the BODIPYs as long-lasting and effective photoactive media of tunable lasers.

2. Light absorption

The reference **PM546** dye displays a sharp and intense (reaching molar absorption at the maximum of 10^5 M^{-1} cm^{-1}) absorption band at around 500 nm (**Figure 2**), showing the trademark low negative solvatochromism (around 7 nm) of BODIPYs [34]. The attachment of linear (ethyl in **PM567**) and branched (*tert*-butyl in **PM597**) alkyls at β-positions (**Figure 1**) induces a spectral bathochromic shift (around 20 and 30 nm, respectively, **Figure 2**), according to an inductive electron releasing effect (higher in *tert*-butyl owing to the hyperconjugative effect). In contrast, the branched substituent decreases the absorption probability (6–7·10^4 M^{-1} cm^{-1}). Such red-shift is more pronounced upon the incorporation of electron acceptor moieties at *meso*-position (**Figure 2**). Thus, the acetoxy moiety at the 8-methyl (**PM605**, **Figure 1**) places the absorption at 550 nm, whereas the stronger electron-withdrawing cyano at such chromophoric position (**PM650**, **Figure 1**) further shifts the absorption to 590 nm (**Figure 2**), together with a decrease of the molar absorption (5–6·10^4 M^{-1} cm^{-1}).

3. Computational chemistry

To understand the substituent-induced spectroscopic changes, we ran theoretical calculations at the density functional theory (DFT) level, in particular using the CAM-B3LYP method and the 6-311 + g* basis set for the geometry optimization, and the time dependent approach (TD DFT) for the simulation of the vertical excitations. TD DFT fails in the description of the energy of the S_1 state but gives a good qualitative simulation of the substituent effect on the energy of the absorption gap. Besides, it is amenable

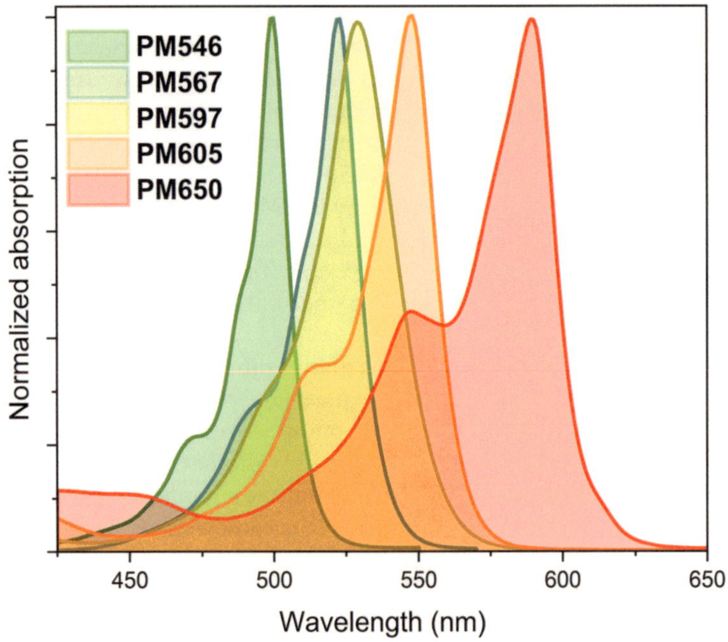

Figure 2.
Normalized absorption spectra of the BODIPY laser dyes in diluted solutions (2 μM) of cyclohexane.

to large molecules, where advance calculation methods are unaffordable [35, 36]. To estimate the TD DFT error and correct the calculated values [37, 38], we compare the TD DFT results of the pristine simplest BODIPY with those calculated with the accurate CASPT2 method [39]. Such comparison reveals that the S_1 energy of BODIPYs is overestimated at around 0.55 eV. Therefore, we applied this correction factor to the absorption energies calculated for the rest of herein studied BODIPY laser dyes using TD DFT. As a result, the predicted absorption energies perfectly match those experimentally recorded (**Figure 3**), supporting the accuracy of TD DFT to reproduce the substituent effects.

Such electronic transition is a result of the promotion of an electron from the HOMO to the LUMO. Therefore, the substituent-induced preferential stabilization/destabilization of these molecular orbitals (OMs) rules the energy gap and the position of the absorption band. On the one hand, the presence of alkyls at positions 2 and 6 raises preferably the energy of the HOMO (**Figure 4**). Indeed, the contribution of these chromophoric positions to the HOMO is slightly higher than in LUMO. Accordingly, electron donor substituents (such as ethyl and *tert*-butyl) destabilize the HOMO with the respect to the LUMO, leading to a slight reduction of the energy gap (from **PM546** to **PM567** and **PM597**, **Figure 4**). On the other hand, electron-withdrawing substituents at position 8 induce a more pronounced decrease in the energy gap owing to a marked reduction of the LUMO energy. In this *meso* position, a node is predicted, being is contribution to the HOMO almost negligible but in contrast very important in the LUMO. Thus, it is a position very sensitive to the substituent effect. Accordingly, the higher the electron acceptor character of the functionalization (acetoxy and cyano) the higher the stabilization of the LUMO (from **PM605** to **PM650, Figure 4**). It is noteworthy that even dim changes in the absorption band position are well reproduced by the theoretically predicted energy of the OMs, being a suitable tool to guide the synthesis of novel BODIPYs with tailored absorption.

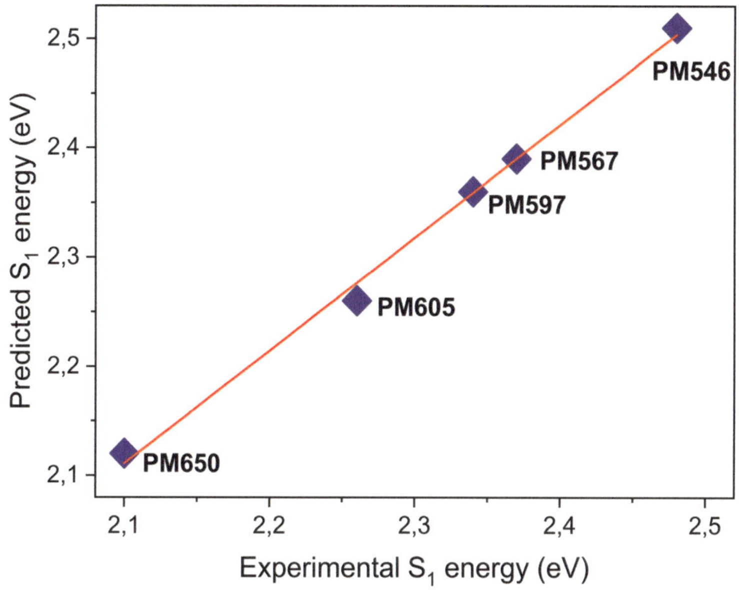

Figure 3.
Correlation between the corrected calculated absorption energies by TD DFT (TD CAM-B3LYP/6-311 + g) and those spectroscopically recorded in cyclohexane.*

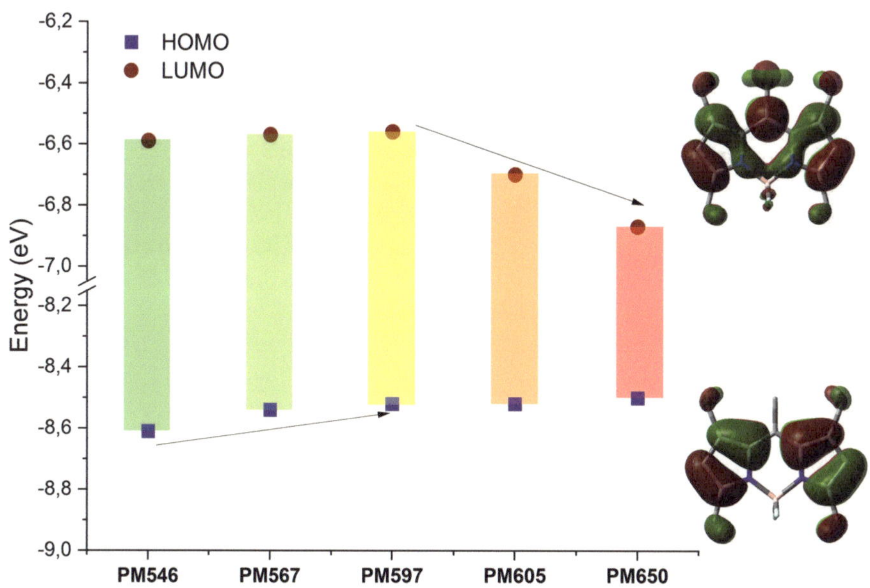

Figure 4.
Representative HOMO and LUMO contour maps and evolution of their energies (in eV) for the studied BODIPYs.

4. Electrochemistry

Further support for the substituent effect in the spectral bands' position is gathered by the redox potentials provided by cyclic voltammetry (**Figure 5**). The simplest BODIPY herein considered (**PM546**) shows two well-defined cathodic and anodic

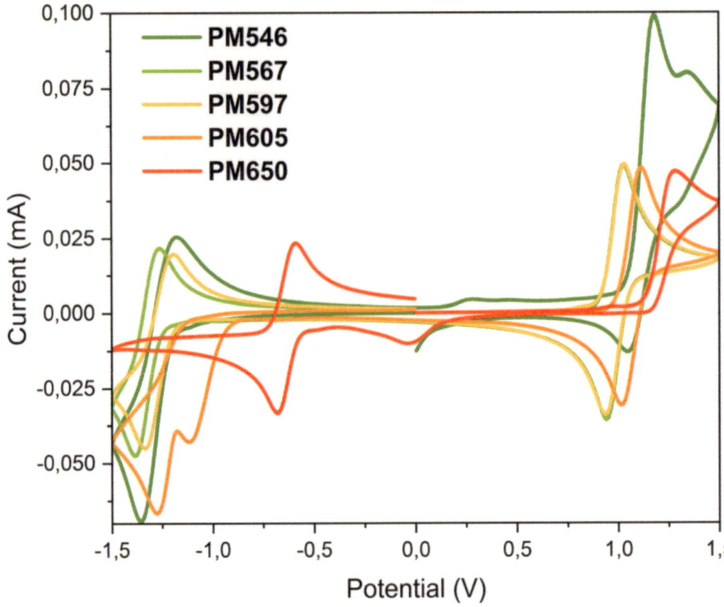

Figure 5.
Cyclic voltammograms in acetonitrile (0.1 m $TBAPF_6$) of the studied BODIPYs (dye concentration 2 mM).

waves at high potentials [40]. The alkylation at positions 2 and 6 makes such waves more reversible and reduces mainly the oxidation potential (from 1.12 V in **PM546** to 0.98 in **PM597**, **Figure 5**). That means, that the ionization potential decreases, in line with the predicted increase of the HOMO energy as the main reason for the recorded bathochromic shift. However, the substitution at position 8 affects mainly the reduction potentials, which become systematically lower as the electron acceptor character increases (from −1.32 V in **PM567** to −1.09 with acetoxy in **PM605** and −0.63 with cyano in **PM650**, **Figure 5**). Thus, these groups increase the electron affinity in agreement with the predicted decrease in the LUMO energy as the source of the recorded large bathochromic shift upon such 8-functionalization.

5. Fluorescence

The evolution of the fluorescence band position with the substituent follows the same trend recorded for the absorption band, but with one exception, **PM597** dye (**Figure 6**). In this case, the bathochromic shift is larger than expected and the profile is broader. Consequently, the Stokes shift is three times higher than in the rest of the dyes (from around 500 to 1500 cm^{-1}, **Table 1**). This feature suggests a marked geometrical rearrangement upon excitation. Another issue to remark is the solvent effect. Most of the dyes show the expected negative solvatochromism, except **PM650** where the opposite positive solvatochromism is detected in polar solvents and just in fluorescence (**Table 1**).

Alkylated dyes, **PM546** and **PM567**, are extremely bright, featuring fluorescence efficiencies close to 100% and long monoexponential lifetimes (around 5–6 ns, **Table 1**). However, in **PM597**, bearing branched alkylation, such efficiency decreases to half. Owing to the quasi-aromatic boron-dipyrrin framework, the spin–orbit

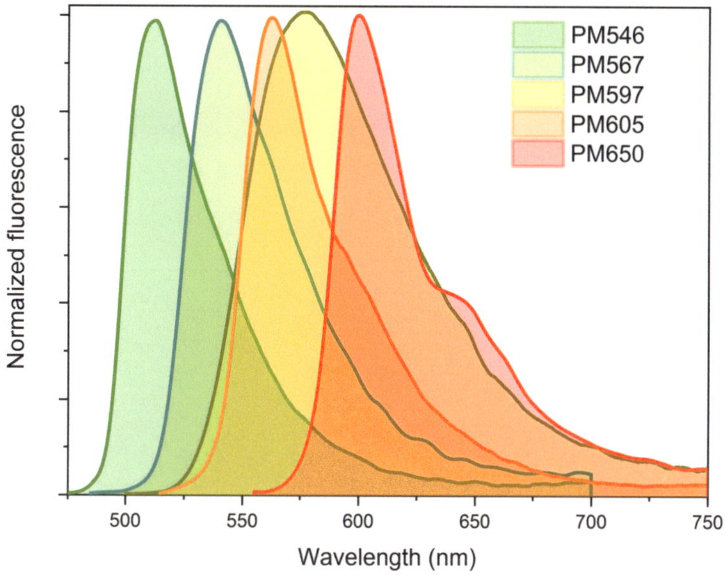

Figure 6.
Normalized fluorescence spectra of the BODIPY laser dyes in diluted solutions (2 μM) of cyclohexane.

coupling is low and the intersystem crossing probability is almost negligible [17]. Thus, the main channel of non-radiative deactivation from the excited state is due to internal conversion, which is related to conformational freedom and planarity [34, 41]. Theoretical calculations predict a chromophoric core essentially planar (slight bending around the transversal axis). Such butterfly-like bending increases in **PM597**, being remarkably higher in the optimized excited state (deviation from planarity up to 15° in the dihedral angle comprising the central ring and the pyrrole). It seems that to accommodate the bulky *tert*-butyl in the peralkylated chromophore and relieve geometrical stress, the planarity is distorted due to steric reasons, and mainly upon excitation, supporting the recorded different bathochromic shift in absorption and fluorescence, and the ensuing high Stokes shift (**Table 1**).

With regard to the *meso* position, the acetoxy group shifts the emission bathochromically, while maintaining a reasonably high fluorescence efficiency (**Table 1**). Compared with the unconstrained alkylated BODIPYs **PM546** and **PM567**, the radiative rate constant of **PM605** is lower and the non-radiative one higher. Likely, the electron-withdrawing effect of the acetoxy retrieves electronic density from the chromophoric core decreasing the aromaticity, as reflected in lower molar absorptions (**Table 1**). Further increase in such electron acceptor character by the attachment of cyano implies more prominent changes in the fluorescence properties. Apart from the mentioned deep bathochromic shift displayed by **PM650** (**Figure 6**), the fluorescence efficiency decreases mainly in polar media (from 50–15%), where the lifetimes are fast (1–2 ns) (**Table 1**). The chromophoric framework of **PM650** is fully planar and the cyano is linear and small, so the fluorescence quenching cannot be assigned to internal conversion as in the constrained alkylated BODIPYs, but rather to the cyano-induced extra non-radiative deactivation pathway, which is favored in polar media. All these trends pinpoint a photoinduced intramolecular charge transfer (ICT) from the BODIPY to the cyano, giving its high electron acceptor ability [42]. In fact, the previously recorded voltammogram for **PM650** reveals a drastic change in the cathodic wave, being this dye

	λ_{ab} (nm)	$\varepsilon_{max} \cdot 10^{-4}$ (M^{-1} cm^{-1})	λ_{fl} (nm)	$\Delta\nu_{St}$ (cm^{-1})	ϕ	τ (ns)	k_{fl} (10^8 s^{-1})	k_{nr} (10^8 s^{-1})	E_{ox} (V)	E_{red} (V)
PM546										
c-hex	499.0	10.3	514.0	490	0.99	5.42	1.82	0.02		
EtOAc	494.0	9.2	506.0	475	0.96	5.58	1.72	0.07		
Acn	492.0	8.7	504.0	490	0.90	5.72	1.57	0.17	1.12	−1.27
PM567										
c-hex	522.5	11.7	541.0	650	0.93	5.98	1.55	0.12		
EtOAc	527.0	8.4	536.0	685	0.83	6.08	1.36	0.28		
Acn	515.0	7.7	534.0	685	0.82	6.31	1.30	0.28	0.98	−1.32
PM597										
c-hex	529.0	7.5	576.0	1545	0.47	4.00	1.17	1.33		
EtOAc	523.0	6.7	564.0	1390	0.48	4.38	1.09	1.18		
Acn	521.0	6.5	563.0	1430	0.51	4.13	1.23	1.19	0.98	−1.26
PM605										
c-hex	547.5	8.2	562.0	470	0.69	6.33	1.09	0.49		
EtOAc	543.0	6.8	560.0	560	0.74	6.71	1.10	0.32		
Acn	541.5	6.6	558.0	550	0.68	6.90	0.98	0.43	1.06	−1.09
PM650										
c-hex	589.0	6.6	598.0	255	0.53	4.73	1.12	1.00		
EtOAc	588.0	6.0	605.0	475	0.50	2.48	0.80	3.22		
Acn	587.5	5.1	608.0	575	0.14	1.67	0.84	8.38	1.28	−0.63

Absorption (λ_{ab}) and fluorescence (λ_{fl}) wavelength, molar absorption at the maximum (ε_{max}), Stokes shift ($\Delta\nu_{St}$), fluorescence quantum yield (ϕ) and lifetime (τ), radiative (k_{fl}) and non-radiative (k_{nr}) rate constants, oxidation (E_{ox}) and reduction (E_{red}) potentials. c-hex: cyclohexane; EtOAc: ethyl acetate; and Acn: acetronitrile.

Table 1.
Photophysical properties of BODIPY laser dyes in diluted solutions (2 μM). Electrochemical data in acetonitrile are also provided.

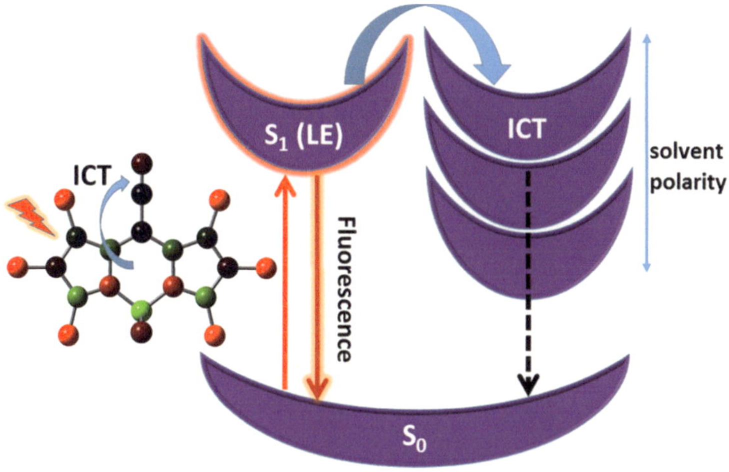

Figure 7.
Sketch of the cyano-induced charge transfer in **PM650** *depending on the solvent polarity. The optimized first excited state geometry with the atoms colored by the charge (positive in red and negative in green) is also included.*

prone to be easily reduced (**Figure 5**). Moreover, the recorded positive solvatochromism exclusively in the fluorescence spectra of **PM650** envisages an increase of the dipole moment upon excitation, according to charge separation processes. ICT states are characterized by high dipole moments owing to the induced charge separation and are further stabilized in polar media. Thus, in these solvents, ICT can become a low-lying state with regard to the locally excited (LE) one, efficiently quenching the fluorescence emission from the latter state (**Figure 7**). This ICT seems to be a dark state, or at least a weakly emissive state, because no new emission bands are recorded, or perhaps the ICT emission is masked under the LE emission. Therefore, the cyano-induced charge separation arises as the non-radiative channel funneling the excited electrons mainly in polar media, where the fluorescence quenching is more notable (**Figure 7**).

6. Laser

Once revisited the fundamental photophysical signatures, we registered the laser properties at high concentrations (millimolar) under transversal pumping by a wavelength-tunable optical parametric oscillator (OPO) coupled to the third harmonic (355 nm) of the Nd:YAG laser. Thus, all the dyes were pumped at their corresponding absorption maximum wavelength. At these conditions, all of them display strong and sharp laser emission signals as expected (**Figure 8**). The laser emission appears within the fluorescence band, but at the end of the long-wavelength absorption tail, where the overlap with the absorption band has vanished, its position nicely correlates with the fluorescence on. Thus, alkylation at positions 2 and 6 induces a bathochromic shift, especially for **PM597**, reinforced upon substitution at position 8 with electron acceptor groups like in **PM650** (**Figure 8**).

However, the correlation of the emission efficiency (laser *vs.* fluorescence) is more complex, and it cannot be done directly, since other factors like the Stokes shift should be considered altogether. We should bear in mind that the laser action is recorded at high optical density media, whereas the photophysical signatures at diluted solutions. At the high concentrations required for the gain overcomes the

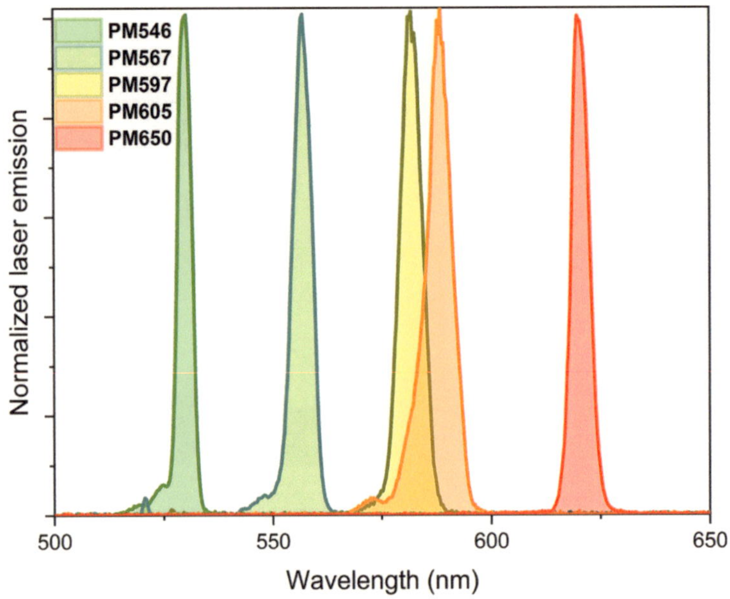

Figure 8.
Laser emission spectra of the BODIPY laser dyes at a high concentration (0.75 mM) in ethyl acetate.

losses in the resonator cavity, other competitive processes can appear. Dye aggregation is discarded owing to the low tendency of these dyes to self-associate in organic media (the spectral profiles remain unaltered even at high concentrations) [34]. However, reabsorption/reemission phenomena are ubiquitous owing to the spectral overlap between the absorption and fluorescence bands (BODIPYs usually feature low Stokes shift). Their influence is notorious in the laser emission wavelength (bathochromic shift of 10–15 nm with the dye concentration), as well as in the laser efficiency since an increase of the dye concentration does not necessarily imply higher efficiency. Instead, a plateau is reached and there is an optimal concentration for each dye, in which the maximum efficiency is recorded (**Figure 9**).

Thus, **PM546** and **PM567** display similar laser efficiencies, albeit slightly higher in the latter (59 vs. 64%, **Figure 9**). Attending to their photophysics, **PM567** is slightly less fluorescence, but it shows a faintly higher Stokes shift, which compensates for such lower efficiency. Further evidence of the key role of the Stokes shift is provided by **PM597**, which displays higher lasing efficiencies (74%, **Figure 9**), albeit its fluorescence efficiency decreases to half (**Table 1**). As mentioned previously, this dye outstands by large Stokes shift (three times that of **PM546**). Therefore, the impact of the reabsorption/reemission is less harmful and counteracts its lower fluorescence yielding very bright laser emission. Indeed, whereas the optimal concentration was 0.5 mM for **PM546** and **PM567**, in **PM597** it is 0.75 mM. Moreover, the laser efficiency drastically decreases for **PM546** (the dye with the shortest Stokes shift, **Table 1**) at high concentrations (1 mM) (**Figure 9**). On the other hand, the substitution at *meso* position has a marked impact on the photophysics (**Table 1**). **PM605** displays reasonably high-lasing efficiencies (up to 66%, **Figure 9**) similar to **PM567**, in concordance with their similar fluorescence efficiencies (even slightly lower in the former, **Table 1**). Finally, **PM650** is the dye endowed with the lowest laser efficiency (54% at the optimal dye concentration, **Figure 9**) as anticipated by the photophysics owing to the ongoing ICT

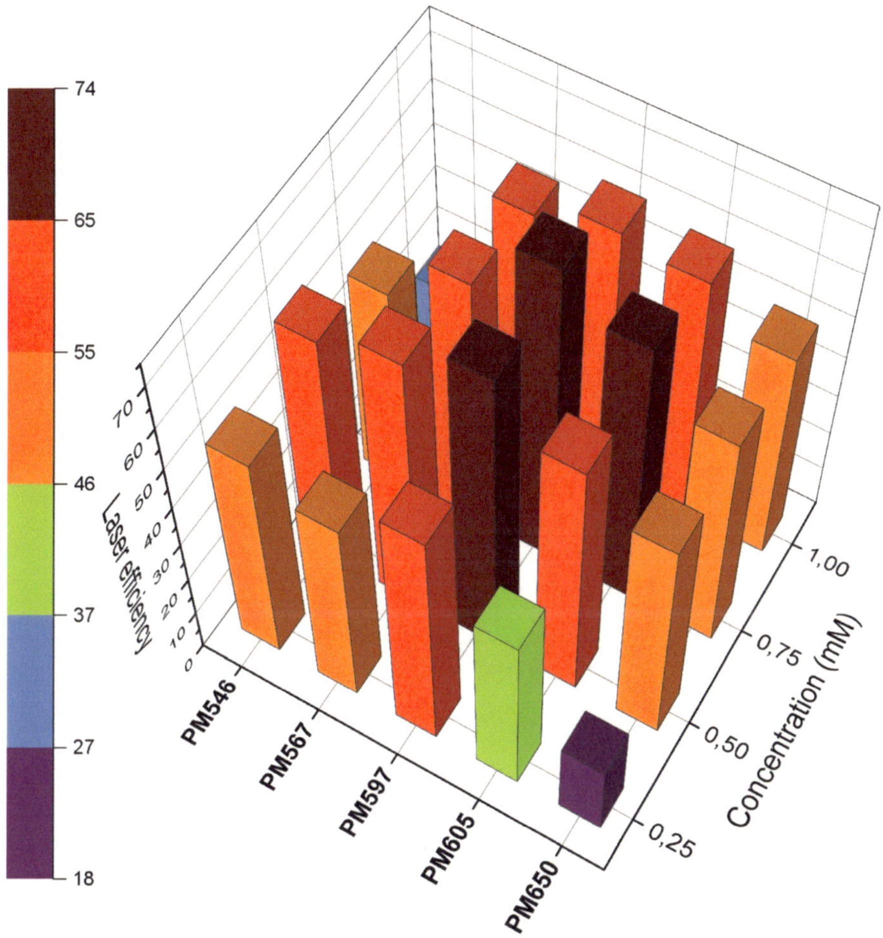

Figure 9.
Laser efficiency of the BODIPY laser dyes at different concentrations in ethyl acetate.

(**Figure 7**). However, the laser performance is much better than envisioned by the low fluorescence efficiency (just a 20%). A remarkable feature of this dye is the shortening of the lifetime (down to 2 ns, **Table 1**), which could enhance the stimulated emission once the population inversion is achieved, and, in this way, counteract partially its low spontaneous emission probability. Therefore, several parameters (fluorescence efficiency, lifetime, and Stokes shift) have to be simultaneously considered to account for the molecular structure effect on the laser output efficiency.

Another critical parameter for the practical implementation of the dyes as active media of lasers is the photostability, in other words, the tolerance to endure strong and continuous irradiation regimes [43]. High photostabilities are required to ensure a long-operative lifetime of the laser. In this regard, the laser emission of **PM546** decreases to half after 50,000 pulses (**Figure 10**). Higher resistance is achieved upon substitution at positions 2 and 6, where **PM597** retains the 80% of its emission after 70,000 pulses (**Figure 10**). Therefore, substitution at the chromophoric positions is recommended to enhance photostability. On the other hand, the substitution at the *meso* position has a marked impact on the photostability. Thus, in **PM605**, the emission completely vanishes after just 40,000 pulses, whereas **PM650** withstands 70,000 pulses with an emission

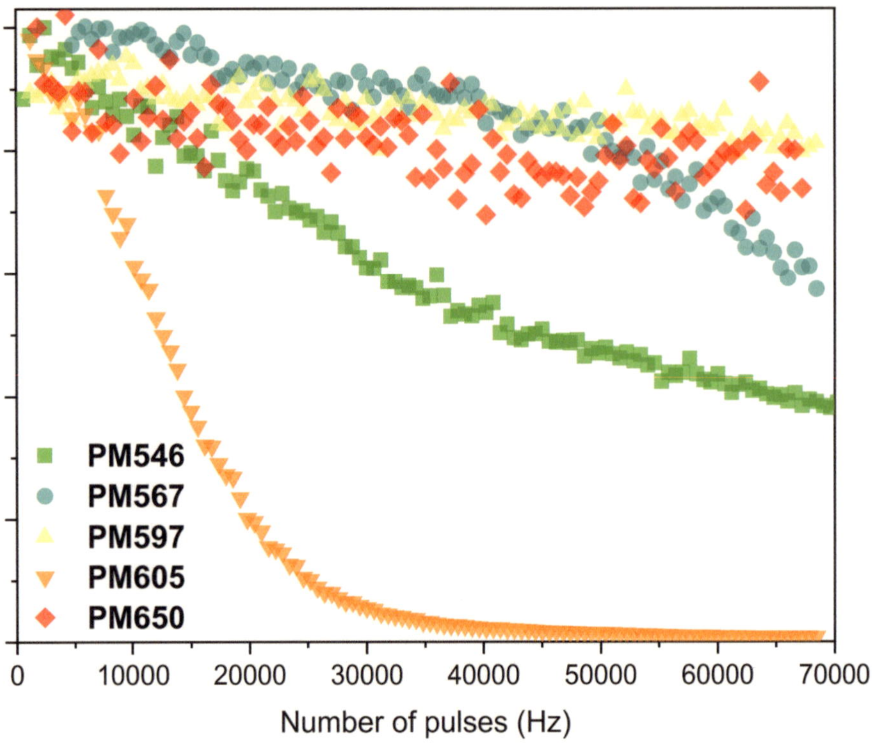

Figure 10.
Photostability of the BODIPY laser dyes under continuous irradiation at dye concentration (0.75 mM) in ethyl acetate.

loss of just 20%. The photodegradation mechanism is not fully unraveled, but it seems to involve the oxidation of the dye by the *in situ*-generated reactive oxygen species (ROS, such as singlet oxygen) upon irradiation, taking place mainly at the *meso* position [44–46]. Thus, steric and electronic reasons can account for the photodegradation rate promoted by oxidative singlet oxygen or related ROS. In **PM650**, the *meso* position is accessible, since the geometry of cyano is linear, and it is more exposed to the singlet oxygen than the rest of dyes bearing 8-methyl. However, this dye is the BODIPY with the highest oxidation potential (1.28 V in **Figure 5**) and hence is less prone to be oxidized by ROS, supporting its drastic increase in photostability owing to electronic reasons (**Figure 10**). Conversely, in **PM605**, the oxidation potential decreases to 1.06 V (**Figure 5**), being easier oxidized. Perhaps, the acetoxy itself contributes also to the generation of ROS under light irradiation and enhances the photooxidation at the *meso* position, leading to cation radicals, and explaining its fast degradation rate (**Figure 10**). Therefore, considering both parameters (efficiency and photostability) alkylated **PM567** and **PM597** are the most recommended laser dyes as powerful and long-lasting photoactive media.

7. Conclusions

We have performed a comprehensive and comparative analysis of the laser performance of commercially available BODIPYs, as well as the photophysical properties, which sustain them. This study highlights that a deep knowledge of the underlying photophysical phenomena, as well as their dependency on the molecular structure, is fundamental to understand the ulterior laser performance. Fully alkylated BODIPYs are suited to yield bright and stable green–yellow lasers, whereas the surprising **PM650** shows good performance as red laser. The interpretation of the relationship between molecular structure and laser behavior is complex, and several photophysical parameters should be considered altogether. For instance, the laser efficiency does not depend solely on the fluorescence efficiency, and other properties, as the Stokes shift and the lifetime, play a key role. Therefore, the combination, which seems to optimize the laser action, endows high fluorescence efficiency and Stokes shift, and fast lifetimes. However, these three properties are antagonisms. Most of the brighter fluorophores are characterized by small Stokes shifts and long lifetimes, while fluorescence quenching usually involves large Stokes shifts and short lifetimes. Therefore, an equilibrium between them should be reached to optimize the laser action.

The herein-reported master lines could serve as a guide to design new fluorophores or modify the available ones to build up benchmark dyes for lasers. In this regard, computational chemistry is a recommended tool to save time and effort in their design, since reproducing nicely the spectral shifts toward the development of cost-effective visible-NIR tunable dye lasers.

Acknowledgements

We acknowledge Spanish MICINN (PID2020-114755GB-C33) and Gobierno Vasco (IT1639-22) for financial support.

Conflict of interest

The authors declare no conflict of interest.

Author details

Alaitz Peñafiel, Ainhoa Oliden-Sánchez, Edurne Avellanal-Zaballa, Leire Gartzia-Rivero, Rebeca Sola-Llano and Jorge Bañuelos-Prieto*
Department of Physical Chemistry, Basque Country University (UPV/EHU), Bilbao, Spain

*Address all correspondence to: jorge.banuelos@ehu.es

References

[1] Sinkeldam RW, Greco NJ, Tor Y. Fluorescent analogs of biomolecular building blocks: Design, properties, and applications. Chemical Reviews. 2010;**110**:2579-2619. DOI: 10.1021/cr900301e

[2] D'Souza RN, Pischel U, Nau WM. Fluorescent dyes and their supramolecular host/guest complexes with macrocycles in aqueous solution. Chemical Reviews. 2011;**111**:7941-7980. DOI: 10.1021/cr200213s

[3] Bessette A. Design, synthesis and photophysical studies of dipyrromethene-based materials: Insights into their applications in organic photovoltaic devices. Chemical Society Reviews. 2014;**43**:3342-3405. DOI: 10.1039/C3CS60411J

[4] Mei J, Leung NLC, Kwok RTK, Lam JWY, Tang BZ. Aggregation-induced emission: Together we shine, united we soar! Chemical Reviews. 2015;**115**:11718-11940. DOI: 10.1021/acs.chemrev.5b00263

[5] Gierschener J, Varghese S, Park SY. Organic single cristal lasers: A materials view. Advanced Optical Materials. 2016;**4**:348-364. DOI: 10.1002/adom.201500531

[6] Kuehne AJC, Gather MC. Organic lasers: Recent developments on materials, device geometries, and fabrication techniques. Chemical Reviews. 2016;**116**:12823-12864. DOI: 10.1021/acs.chemrev.6b00172

[7] Zhang W, Yao J, Zhao YS. Organic micro/nanoscale lasers. Accounts of Chemical Research. 2016;**49**:1691-1700. DOI: 10.1021/acs.accounts.6b00209

[8] Mirkovic T, Ostroumov EE, Anna JM, Van Grondelle R, Govindjee SGD. Light absorption and energy transfer in the antenna complexes of photosynthetic organisms. Chemical Reviews. 2017;**117**:249-193. DOI: 10.1021/acs.chemrev.6b00002

[9] Li X, Kolemen S, Yoon J, Akkaya EU. Activatable photosensitizers: Agents for selective photodynamic therapy. Advanced Functional Materials. 2017;**27**:1604053. DOI: 10.1002/adfm.201604053

[10] Wu D, Sedgwick AC, Gunnlaugsson T, Akkaya EU, Yoon J, James TD. Fluorescent chemosensors; the past, present and future. Chemical Society Reviews. 2017;**46**:7105-7123. DOI: 10.1039/C7CS00240H

[11] Nguyen VN, Yan Y, Zhao J, Yoon J. Heavy-atom-free photosensitizers: From molecular design to applications in the photodynamic therapy of cancer. Accounts of Chemical Research. 2021;**54**:207-220. DOI: 10.1021/acs.accounts.0c00606

[12] De Moliner F, Kielland N, Lavilla R, Vendrell M. Modern synthetic avenues for the preparation of functional fluorophores. Angewandte Chemie, International Edition. 2017;**16**:3758-3769. DOI: 10.1002/anie.201609394

[13] Kulyk O, Rocard L, Maggini L, Bonifazi D. Synthetic strategy tailoring colours in multichromophoric organic nanostructures. Chemical Society Reviews. 2020;**49**:8400-8424. DOI: 10.1039/C9CS00555B

[14] Loudet A, Burgess K. BODIPY dyes and their derivatives: Syntheses and spectroscopic properties. Chemical

Reviews. 2007;**107**:4891-1932. DOI: 10.1021/cr078381n

[15] Boens N, Verbelen B, Ortiz MJ, Jiao L, Dehaen W. Synthesis of BODIPY dyes through postfunctionalization of the boron dipyrromethene core. Coordination Chemistry Reviews. 2019;**399**:213024. DOI: 10.1016/j.ccr.2019.213024

[16] Poddar M, Misra R. Recent advances of BODIPY based derivatives for optoelectronic aplications. Coordination Chemistry Reviews. 2020;**421**:213462. DOI: 10.1016/j.ccr.2020.213462

[17] Pavlopoulos TG. Scaling of dye lasers with improved laser dyes. Progess in Quantum Electronics. 2002;**26**:193-224. DOI: 10.1016/S0079-6727(02)00005-8

[18] Treibs A, Kreuzer FH. Justus Liebigs Ann. Chem. 1968;**718**: 208-223

[19] Shah M, Thangaraj K, Soong ML, Wolford LT, Boyer JH, Politzer IR, et al. Pyrromethene-BF_2 complexes as laser dyes: 1. Heteroatom Chemistry. 1990;**1**:389-399. DOI: 10.1002/hc.520010507

[20] Boyer JH, Haag AM, Sathyamoorthi G, Soong ML, Thangaraj K, Pavlopoulos TG. Pyrromethene-BF_2 complexes as laser dyes: 2. Heteroatom Chemistry. 1993;**4**:39-48. DOI: 10.1002/hc.520040107

[21] Boens N, Leen V, Dehaen W. Fluorescent indicators based on BODIPY. Chemical Society Reviews. 2012;**41**:1130-1172. DOI: 10.1039/C1CS15132K

[22] Klfout H, Stewart A, Elkhalifa M, He H. BODIPYs for dye-sensitized solar cells. ACS Applied Materials & Interfaces. 2017;**9**:39873-39889. DOI: 10.1021/acsami.7b07688

[23] Kue CS, Ng SY, Voon SH, Kamkaew A, Chung LP, Kiew LV, et al. Recent strategies to improve boron dipyrromethene (BODIPY) for photodynamic cancer therapy: An updated review. Photochemical & Photobiological Sciences. 2018;**17**:1691-1708. DOI: 10.1039/C8PP00113H

[24] Agazzi ML, Bellatore MB, Durantini AM, Durantini EN, Tomé AC. BODIPYs in antitumoral and antimicrobial photodynamic therapy: An integrating review. Journal of Photochemistry and Photobiology C. 2019;**40**:21-48. DOI: 10.1016/j.jphotochemrev.2019.04.001v

[25] Ma JL, Peng Q, Zhao CH. Circularly polarized luminescence switching in small organic molecules. Chemistry - A European Journal. 2019;**25**:15441-15454. DOI: 10.1002/chem.201903252

[26] Kaur P, Singh K. Recent advances in the application of BODIPY in bioimaging and chemosensing. Journal of Materials Chemistry C. 2019;**7**:11361-11405. DOI: 10.1039/C9TC03719E

[27] Shi Z, Han X, Hu W, Bai H, Peng B, Ji L, et al. Bioapplications of small molecule Aza-BODIPY: From rational structural design to in vivo investigations. Chemical Society Reviews. 2020;**49**:7533-4567. DOI: 10.1039/D0CS00234H

[28] De Bonfils P, Péault L, Nun P, Coeffard V. State of the art of bodipy-based photocatalysts in organic synthesis. European Journal of Organic Chemistry. 2021;**2021**:1809-1824. DOI: 10.1002/ejoc.202001446. Available from: https://chemistry-europe.onlinelibrary.wiley.com/doi/abs/10.1002/ejoc.202001446

[29] Lu H, Mack J, Yang Y, Shen Z. Structural modification strategies for the rational design of red/NIR region BODIPYs. Chemical Society Reviews.

2014;**43**:4778-4823. DOI: 10.1039/C4CS00030G

[30] Yariv E, Schultheiss S, Saraidarov T, Reisfeld R. Efficiency and photostability of dye-doped solid-state lasers in different hosts. Optical Materials. 2001;**16**:29-38. DOI: 10.1016/S0925-3467(00)00056-2

[31] Ahmad M, King TA, Ko DK, Cha BH, Lee J. Performance and photostability of xanthene and pyrromethene laser dyes in sol-gel phases. Journal of Physics D. 2002;**35**:1473-1476. DOI: 10.1088/0022-3727/35/13/303

[32] Costela A, Garcia-Moreno I, Sastre R. Polymeric solid-state dye lasers: Recent developments. Physical Chemistry Chemical Physics. 2003;**5**:4745-4763. DOI: 10.1039/B307700B

[33] Chénais S, Forget S. Recent advances in solid-state organic lasers. Polymer International. 2012;**61**:390-406. DOI: 10.1002/pi.3173

[34] Bañuelos J. BODIPY dye, the most versatile fluorophore ever? Chemical Record. 2016;**16**:335-348. DOI: 10.1002/tcr.201500238

[35] Laurent AD, Adamo C, Jacquemin D. Dye chemistry with time-dependent density functional theory. Physical Chemistry Chemical Physics. 2014;**16**:14334-14336. DOI: 10.1039/C3CP55336A

[36] Momeni MR, Brown A. Why do TD-DFT excitation energies of BODIPY/Aza-BODIPY families largely deviates from experiments? Answers from electron correlated and multireference methods. Journal of Chemical Theory and Computation. 2015;**11**:2619-2632. DOI: 10.1021/ct500775r

[37] Schlachter A, Fleury A, Tanner K, Soldera A, Habermeyer B, Guilard R, et al. The TDDFT excitation energies of the BODIPYs; the DFT and TDDFT challenge continues. Molecules. 2021;**26**:1780. DOI: 10.3390/molecules26061780

[38] Postils V, Ruipérez F, Casanova D. Mild open-shell character of BODIPY and its impact on singlet and triplet excitation energies. Journal of Chemical Theory and Computation. 2021;**17**:5825-5838. DOI: 10.1021/acs.jctc.1c00544

[39] De Vetta M, González L, Corral I. The role of electronic triplet states and high-lying singlet states in the deactivation mechanism of the parent BODIPY: An ADC(2) adn CASPT2 study. ChemPhotoChem. 2019;**3**:727-738. DOI: DOI. 10.1002/cptc.201800169

[40] Nepomnyashchii AB, Bard AJ. Electrochemistry and electrogenerated chemiluminescence of BODIPY dyes. Accounts of Chemical Research. 2012;**45**:1844-1853. DOI: 10.1021/ar200278b

[41] Lin Z, Kohn AW, Van Voorhis T. Toward prediction of nonradiative decay pathways in organic compounds II: Two internal conversion channels in BODIPYs. Journal of Physical Chemistry C. 2020;**124**:3925-3938. DOI: 10.1021/acs.jpcc.9b08292

[42] López Arbeloa F, Bañuelos J, Martínez V, Arbeloa T, López AI. Structural, photophysical and lasing properties of pyrromethene dyes. International Reviews in Physical Chemistry. 2005;**24**:339-374. DOI: 10.1080/01442350500270551

[43] Demchenko AP. Photobleaching of organic fluorophores: Quantitative characterization, mechanisms, protection. Methods and Applications in Fluorescence. 2020;**8**:022001. DOI: 10.1088/2050-6120/ab7365

[44] Jones G II, Kumar S, Klueva O, Pacheco D. Photoinduced electron transfer for pyrromethene dyes. The Journal of Physical Chemistry. A. 2003;**107**:8429-8434. DOI: 10.1021/jp0340212

[45] Cui A, Peng X, Fan J, Chen X, Wu Y, Guo B. Synthesis, spectral properties and photostability of novel boron-dipyrromethene dyes. Journal of Photochemistry and Photobiology A. 2007;**186**:85-92. DOI: 10.1016/j.jphotochem.2006.07.015

[46] Mula S, Ray AK, Banerjee M, Chaudhuri T, Dasgupta K, Chattopadhyay S. Design and development of a new pyrromethen dye with improved phostostability and lasing efficiency: Theoretical rationalization of photophysical and photochemical properties. The Journal of Organic Chemistry. 2008;**73**:2146-2154. DOI: 10.1021/jo702346s

Chapter 2

Green Synthesis of TiO_2 Nanoparticles Using Averrhoa Bilimbi Fruits Extract and DPT-PEG Polymer Electrolyte for Enhance Dye-Sensitized Solar Cell Application

Sundaramurthy Devikala and Johnson Maryleedarani Abisharani

Abstract

Green synthesis of nanoparticles has grown substantial interest as a developing technology to reduce the toxicity of metal oxide commonly associated with conventional physical and chemical synthesis methods. Among these, green synthesis of nanoparticles from plants parts to be a very active method in developing nontoxic, eco-friendly and clean technology. We prepared green synthesized TiO_2 using a fruits extract of *Averrhoa bilimbi* with a cost effective and non-toxic method and reports better PCE of DSSCs application. The green synthesized TiO_2 nanoparticles (working electrode) with DPT dopant PEG polymer electrolyte shows better power conversion efficiency in dye-sensitized solar cells. The green TiO_2 was characterized with XRD, UV, FTIR, SEM, TEM and EDX techniques analysis the band gap, crystallite size and shape for green synthesized TiO_2 nanoparticles. The electrical and mechanical properties of DPT organic doped $PEG/KI/I_2$ polymer electrolyte were characterized with XRD, FTIR, EIS, DSC and TGA and it was analysis that the DPT well miscible with PEG polymer electrolyte and improves the electrical conductivity and enhances the efficiency of DSSC.

Keywords: *Averrhoa bilimbi*, green synthesis, titanium dioxide, DPT organic compound, DSSCs

1. Introduction

Dye-Sensitized Solar Cells (DSSCs) have attracted researchers owing to their unexpected potential of lightweight, inexpensive cost materials, flexible structure, and easy fabrication. Substantial researches have been conducted for the progress of Dye sensitized solar cells but the development of DSSCs is still not feasible. Although the PCE of the DSSCs are lower while comparing the first and second-generation PV (photovoltaic) cells and there still the development of a high potential for significant efficiency [1, 2]. Vogel et al. in 1870 proposed a first DSSCs by using silver halide, but the device could not show any output. In 1887 James Moser, developed initial voltage was around 0.04 V in DSSCs. Hishiki and Gerischer (1965–1968) introducing ZnO and sensitizer (Rose Bengal and Cyanine). In 1977, ZnO was replaced with TiO_2 by Spitler & Calvin group and they were clarified two factors, which are the dye molecules absorbed on the TiO_2 surface and pH solution used for the dye absorption process. After many years, these cells were developed in 1991 by Michael Gratzel and Brian O′ Regan, they achieved a PCE of 7.1%. The use of TiO_2, Zinc porphyrin dye and $Co^{2+/3+}$ tris (bipyridine) based electrolyte reached a PCE of 12.3%. Recently, DSSCs achieved a PCE 14.3% efficiency by using two metal-free organic dyes. TiO_2 coated on conductive substrate (photo anode), sensitizer (dye), electrolyte (redox couple) and counter electrode (platinum) [3, 4]. The photo-anode has a wide bandgap of metal oxides such as ZnO, ZrO_2, CuO and SnO_2 etc. commonly used to deposit on the surface of the TCO glass substrate and is used in FTO glass. The dye molecules (sensitizer) are usually absorbed on the surface of the metal oxides. The main components of the dye are part of the DSSCs, which creates light harvesting and produce photoexcited electrons. Electrolyte in DSSCs is used to transfer electrons between both electrodes and regenerate of oxidized dye [5]. The counter electrode is an important component and promotes high catalytic activity owing to the redox pair reaction by electron recombination process, requires high electron transport to improve the conductivity and high reflectance cell assembly to divert the unabsorbed light energy and improve sunlight capture [6].

1.1 Nanocrystalline TiO_2

The nanocrystalline TiO_2 (Titanium Dioxide) is used in several applications such as pigments, photocatalytic, paints, photovoltaic and antimicrobial activities [7–11]. TiO_2 plays an important role in dye sensitized solar cells, because of its small particle size, high surface area, highly active anatase phase, low density, high electron mobility and high band gap energy [12–15]. TiO_2 is an n-type semiconductor material with a 3.2 eV band gap energy. It has three natural mineral forms in the earth that are rutile, anatase and brookite. The most stable form among them is rutile, which is in the equilibrium phase at any temperature. When used in dye-sensitized solar cells, anatase is perceived to be more chemically active than rutile form. Anatase is a metastable, and when heated, it tends to convert to rutile. As a result, the synthesis has a significant impact on the phase constituents. At approximately 25 nm, the commercial product DeGussa P25 contains 80% anatase and 20% rutile. The simulated AM 1.5 solar illumination was used to test dye-sensitized solar cells (DSSCs) made with rutile and anatase films of the same thickness. The results showed that the open-circuit voltage (V_{oc}) is essentially the same, but the short-circuit photocurrent (I_{sc}) of the anatase-based cell is 30% higher than that of the rutile-based cell. The variance in short circuit current is attributed to the rutile film's lower dye absorption due to a smaller specific surface area [14, 16].

Generally, TiO_2 nanoparticles were prepared using a different type of physical and chemical methods, including microwave method, chemical vapor deposition, solvothermal, hydrothermal, sonochemical method, sol–gel, and electrophoretic deposition [17–23]. All of these methods are very expensive, require a lot of pressure, a lot of energy, and pollute the environment, whereas the green synthetic method has been found to be more advantageous than the other methods reported. Plant extracts and plant parts such as leaves, fruits, flowers, seeds, yeast, fungi, bacteria and algae were used in the green synthesis method to produce nanoparticles in the nano range (10–100 nm) [24, 25]. The plant extract supports in the reduction and capping of agents on the surface of nanoparticles. In 2011, Nyctanthes arbor-tristis was used to synthesize TiO_2 using a green synthetic method [13, 26]. *A. bilimbi* is a medicinal plant that is used to treat diabetes, hypertension, and acts as an anticancer and antimicrobial agent [27]. The plant extract contains flavonoids, phenol and tannins found out by phytochemical studies. This functional compound can play as a good stabilizing and capping agent for the formation of nanoparticles [28–30].

1.2 Organic compound dopant electrolytes

The polymer-based electrolytes like poly-urethane (PU), poly (ethylene oxide) (PEO), poly (vinylidene difluoride) (PVDF), poly (vinyl chloride) (PVC), poly acrylonitrile (PAN), etc., were used for DSSCs applications to improve the efficiency and stability [31–36]. Additionally, researchers improve the electrolyte stability using organic dopant as an additive molecule such as alkylaminopyridine, pyridine, alkylpyridine, pyrzaole, benzimidazole, triphenylamine, quinoline, etc. The addition of small number of organic molecules changed the the redox potential, surface of the semiconductor, TiO_2 conduction band edge shift and change the recombination kinetics [37]. Boschloo et al. found the additives affect the TiO_2 semiconductor. These effects, increase of V_{oc} is due to a combination of TiO_2 conduction band edge change toward negative potentials and improve the electron lifetime [38]. As a polymer gel electrolyte content, phenanthroline-based cobalt redox couple and thiourea derivatives were used as additives with hydroxyl propyl cellulose polymer host, resulting in a 9.1% performance. The most stable thioureas have more electron donation groups and have high adsorption energy toward TiO_2. The Fermi level of TiO_2 (101) anatase was shifted by these additives. Furthermore, the performance of the Cobalt redox pair was increased due to its predominating properties [39]. Omid et al., introduced new inexpensive propyl isonicotinate and isopropyl isonicotinate by pyridine derivative as an additive in bromide/tribromide electrolytes revealed PCE 2.81–3.76% [40]. Ganesan et al., investigated the PVdF-PEO/KI/I_2 polymer mixture electrolyte system with DPA (Diphenyl amine) and PT (Phenothiazine) obtained an PCE 8.5% [32].

2. Experimental

2.1 Materials

A. bilimbi fruits were collected from Kanyakumari district, Tamilnadu, India. TiO (SO_4), 2,4-Diamino-6-Phenyl-1-3-5-Triazine, PEG, KI, I_2 and DMF solvent purchased from sigma Aldrich. All the above chemicals were used without further purification.

2.1.1 Preparation of TiO_2 nanoparticles

First the fruits were cleaned thoroughly 2–3 times using distilled water and cut into small pieces. Then, 50 g of the fruits was added in DI water (100 ml) followed by heated for at 1 h. The above mixture filtrated by using Whatman filter paper (125 mm). Filtered fruits extract was collected and stored in refrigeration for using further usages.

Furthermore, 10 ml of fruits extract was taken along with $TiO(SO_4)$ (8 ml) dissolved in DI water (100 ml) and stirred well using a magnetic stirrer with high rpm. After 4 h, the white precipitate gradually formed. Finally, the precipitates were centrifuged for 3–4 times using distilled water. The colloidal mixture was dried at 100°C and calcinated at 400°C and the nanoparticles were collected.

2.1.2 Preparation of DPT polymer electrolyte

Poly(ethylene glycol) (300 mg), 2,4-Diamino-6-Phenyl-1-3-5-Triazine (10 mg), Iodine (10 mg) and Potassium iodide (30 mg) were dissolved in 3 ml of DMF solvent. The electrolyte mixture was continuously stirred at 60°C for 3 h and electrolyte used in DSSC.

2.1.3 DSSCs cell fabrication

The prepared TiO_2 (working electrode) was coated by the doctor blade method as previous literature [41]. The TiO_2 coated on FTO (Fluorinated tin oxide) were immersed in a N3 dye (5×10^{-4} M dye) solution in ethanol for 24 hrs. Then, dye coated TiO_2 plate was dried and used for the measurement of conversion of solar energy to electrical energy. Preparation of electrolyte, PEG (300 mg), KI (30 g), Iodine (10 mg) and 2,4-Diamino-6-Phenyl-1-3-5-Triazine (10 mg) dissolved in DMF solvent. A sandwich type of DSSC cell consisting of N_3 dye-coated TiO_2 and Platinum coated on FTO was used. Then, prepared iodine electrolyte (I^-/I_3^-) solutions were placed in between the N_3 dye coated TiO_2 and Pt electrodes. Finally, the fabricated DSSC are measured in current–voltage (I–V) under sunlight.

2.2 Characterization

Prepared TiO_2 nanoparticles characterize the phase form and crystal size using XRD (Malvern Panalytical). Optical properties of prepared TiO_2 nanoparticles were analyzed by UV–Visible spectroscopy (Agilent Cary 5000). The functional groups present in the prepared TiO_2 nanoparticles were recorded by FTIR using BRUKERα-E (ATR, Lab India Instruments Pvt. Ltd), and the absorbed range of 400–1000 cm^{-1}. Surface morphological and element compositions were observed by SEM and EDX (Quanta 200, FEI). Size of the TiO_2 nanoparticles were measured using TEM (JEOL, TEM-2100 plus electron microscopy, made in Japan). Conductivity study was recorded by Biologic SP-300. Photo electrochemical (IV) properties were studied under the sun illumination of 100 $mWcm^{-2}$ at AM 1.5. The I–V curve was measured using a BAS 100A electrochemical analyzer. The DSSC active cell area was 1 cm^2 (1 cm x 1 cm).

3. Results and discussions

3.1 Characterization for green synthesized TiO_2 nanoparticles

UV–Visible absorption spectroscopy was used to examine the optical properties of the GS-TiO_2 nanoparticles. The light absorption characteristics of GS-TiO_2 nanoparticles are shown in **Figure 1a**. GS-TiO_2 nanoparticles showed a strong absorption peak at 320 nm in the UV–Visible spectrum. This value matches well with the

DOI: http://dx.doi.org/10.5772/intechopen.106944

Com-TiO_2 (Commercial TiO_2) reported in the literature [11]. The optical band-gap of the nanoparticles was calculated using the following equation,

$$(\alpha h\nu)^{\frac{1}{2}} = \beta\left(h\nu - E_g\right) \tag{1}$$

Where, β-constant, α-absorption coefficient (cm^{-1}), E_g-band gap of material, ν-photon frequency, h-Planck's constant.

The optical band gap energy (**Figure 1b**) derived from Tauc's plot corresponds to 3.2 ev (Indirect band gap) good agreement with the anatase phase of Com-TiO_2, which is found to be GS-TiO_2 nanoparticles has good band gap energy of semiconductor materials for photoanode in DSSCs application.

X-ray diffraction spectroscopy was identified crystallite structure and phase formation of GS-TiO_2 nanoparticle by using *Averrhoa bilimbi* fruits extract. **Figure 2 (a)** showed the XRD pattern of GS-TiO_2 nanoparticles. From the XRD pattern, displayed diffraction peaks at 25.3° (101), 37.8° (004), 47.9° (200), 53.9° (105), 55.1° (211), 62.5° (204), 68.9° (116), 74.0° (107) which corresponds to tetragonal structure. The crystallographic plane 101 indicated the anatase phase structure

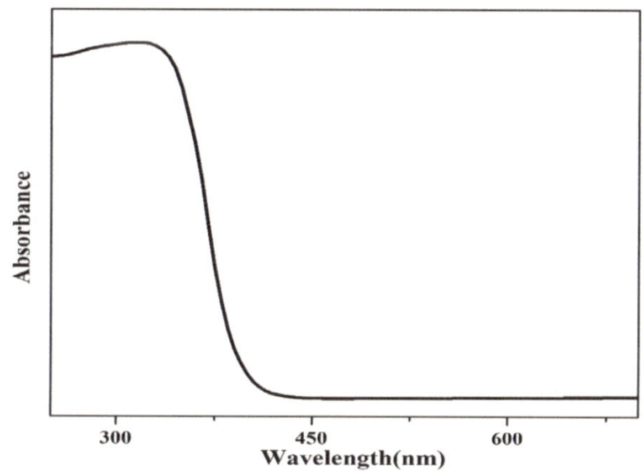

(a)

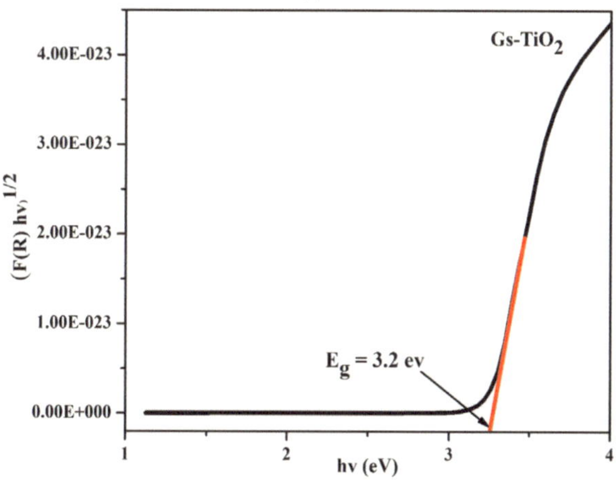

(b)

Figure 1.
(a) UV–visible spectrum of GS-TiO_2 nanoparticles. (b) Band gap of GS-TiO_2 nanoparticles.

formation of GS-TiO_2 was matched with ref. No. 01-084-1285 and that the GS-TiO_2 nanoparticles showed highly crystalline and pure phase formation. Com-TiO_2 XRD pattern showed in **Figure 2(b)**. The average crystalline size of GS-TiO_2 nanoparticles was calculated using Scherrer's formula [42]. The estimated average crystallite size was measured by using the first 3 major peaks of GS-TiO_2 nanoparticles was found to be 23.8 nm based on FWHM (full width at half-maximum) from the XRD pattern.

$$D = \frac{k\lambda}{\beta cos\theta} \tag{2}$$

Where, D—average crystallite size, λ—wavelength, k—Debye Scherrer constant, θ—Bragg's diffraction angle, β—full width half-maximum (FWHM).

The FTIR spectrum of GS-TiO_2 and *Averrhoa bilimbi* fruits extract were show in **Figure 3(a)** and **(b)** respectively. The transmittance was observed in the 4000–500 cm^{-1} range. The *A. bilimbi* fruits extract was analyzed using FTIR spectroscopy to see if any organic functional groups could act as a capping agent on the GS-TiO_2 nanoparticles. The peaks (**Figure 3(a)** at the frequencies 3440, 2926, 1635,

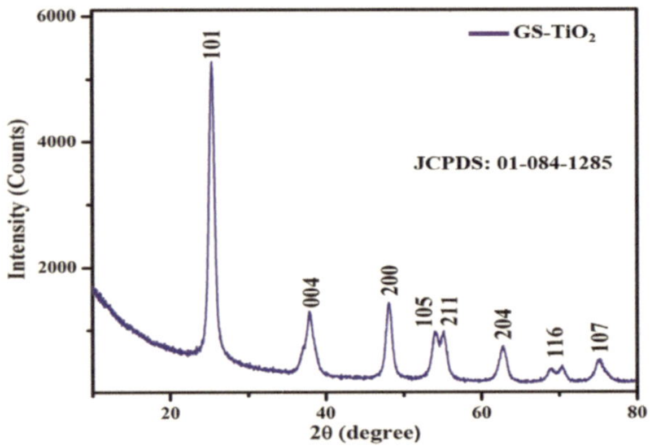

(a)

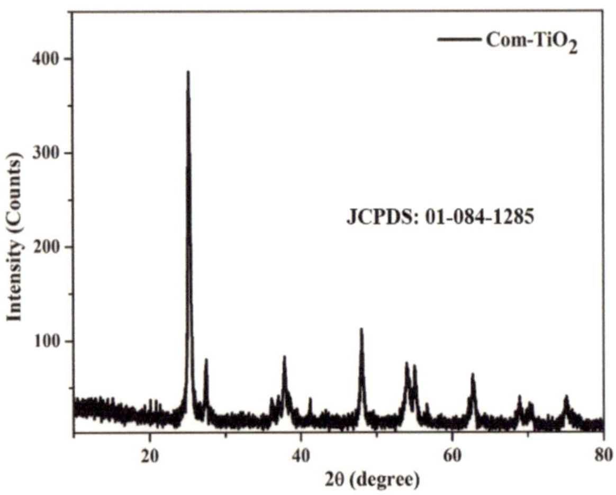

(b)

Figure 2.
(a) XRD pattern of GS-TiO_2 nanoparticles. (b) XRD pattern of Com-TiO_2 nanoparticles.

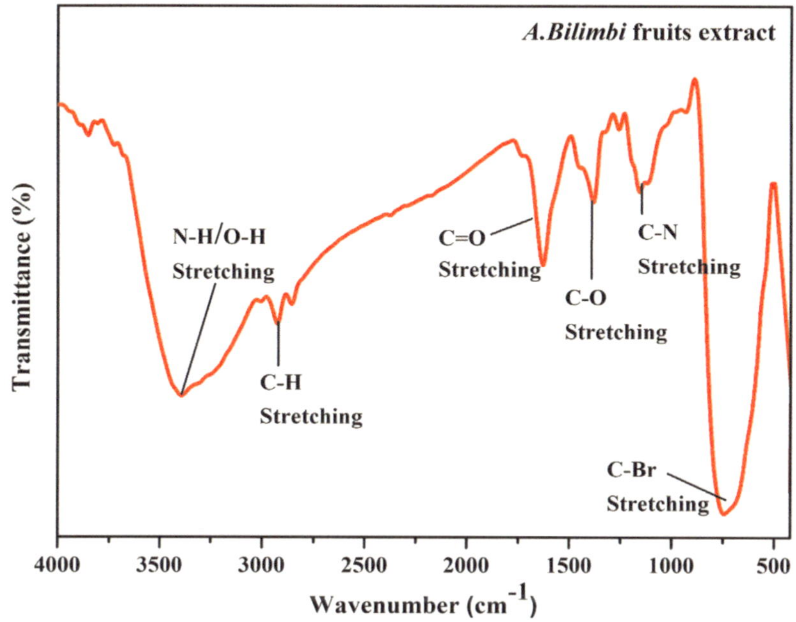

(a)

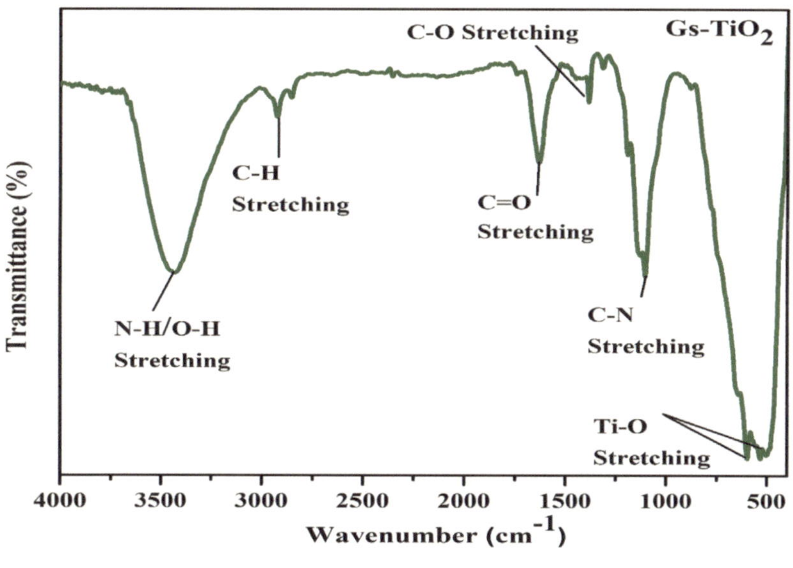

(b)

Figure 3.
(a) FTIR spectrum of Averrhoa bilimbi *fruits extract. (b) FTIR spectrum of GS-TiO_2 nanoparticles.*

1382, 1104, 750 cm^{-1} corresponded to the NH and OH group, C—H stretching and bending, C=O stretching, C—O stretching, C—N stretching and C-Br stretching frequency respectively [43]. In **Figure 3(b)**, Ti-O stretching modes are represented by the absorption peak between 500 and 600 cm^{-1} [44]. From the **Figure 3(a)** and **(b)** showed the FTIR spectroscopy of GS-TiO_2 and *A. bilimbi* fruits extract similar vibration frequency. The stretching frequency of alkyl halide peak was shifted from

750 to 600 cm^{-1}.This demonstrated that *A. bilimbi* fruits extract acts as a capping layer on the GS-TiO_2 nanoparticles.

Morphology and particle size of the GS-TiO_2 nanoparticles were characterized by SEM and TEM analyses. **Figure 4** showed the GS-TiO_2 nanoparticles was agglomerated and uneven spherical shape. TEM images (**Figure 5**) revealed that similar morphology structure with uneven particle size and highly crystallized with average particle size was found to be15 nm of GS-TiO_2 nanoparticles, respectively. The anatase phase form of GS-TiO_2 nanoparticles tetragonal lattice planes (101) was matched to the lattice spacing of 0.35 nm, which were obtained from XRD pattern [45].

Overall, the SEM and TEM analyses demonstrated that the green synthesis method was very useful for the preparation of nano sized, pure anatase phase and mesoporous structural properties of GS-TiO_2 nanoparticles. Because of the developed GS-TiO_2 has the greatest potential for charge transfer process and dye adsorption on the pure anatase surface of TiO_2, it could be used to develop an efficient DSSCs device.

The EDX analysis (**Figure 6**) showed the purity of the GS-TiO_2 nanoparticles as well as the presence of Ti and O at atomic percentages of 45.71 and 54.29, respectively. Because no other peaks determined in the EDX spectrum, the purity of the GS-TiO_2 nanoparticles was clearly shown.

3.1.1 Characterization of DPT doped polymer electrolyte

DSC analysis was used to study about the thermal stability of the polymer electrolytes. The DSC curves of the polymer electrolyte measurement range was set between 40°C and 300°C in nitrogen atmosphere. As shown in **Figure 7**, the melting temperature of prepared polymer electrolyte (PEG/KI/I_2) observed a sharp endothermic peak at 58°C and after DPT doped in electrolyte exhibited more broadening endothermic peak at 86°C, respectively. This broadening endothermic peak confirms the DPT organic compound well interacted with the PEG/KI/I_2 electrolyte and improved the conductivity in PEG polymer.

The weight loss of the PEG polymer electrolyte was determined using thermogravimetric analysis. **Figure 8** depicts the PEG polymer electrolyte weight loss at 420°C with the scan range set between 50 and 500°C [46]. All the electrolyte showed the first stage of weight loss from 100 to 180°C, due to the decomposition of PEG polymer matrix. The

Figure 4.
SEM images of GS-TiO_2 nanoparticles.

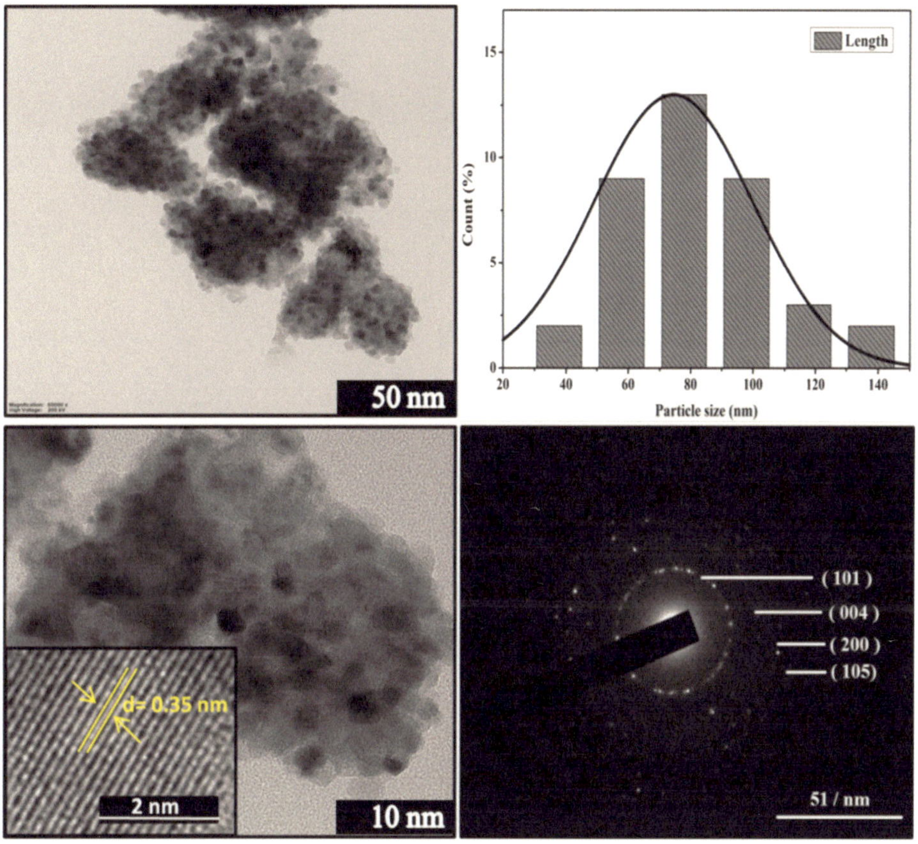

Figure 5.
TEM images of GS-TiO_2 nanoparticles.

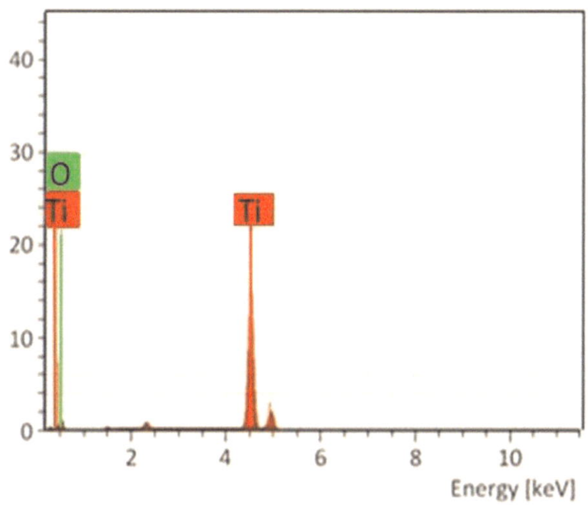

Figure 6.
EDX spectrum of GS-TiO_2 nanoparticles.

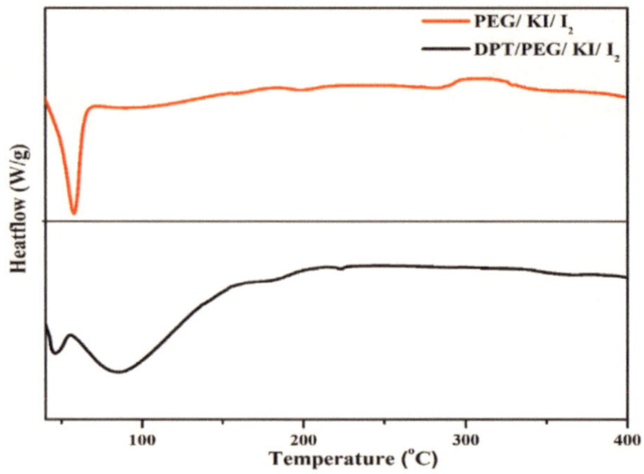

Figure 7.
DSC analysis of polymer electrolytes.

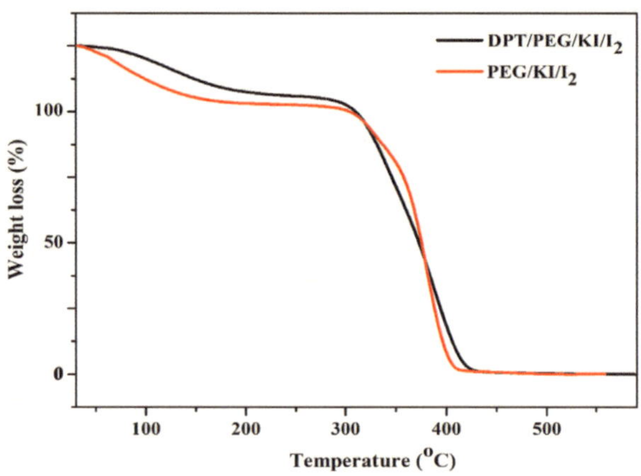

Figure 8.
TGA analysis of polymer electrolyte.

TGA graph showed that the DPT interacts well with PEG and maintains thermal stability, confirming that the synthesized organic compound doped polymer electrolyte decomposed above 420°C. Altogether, both thermal stability changes in the melting temperature and weight loss range can be observed, after the addition of DPT doped PEG polymer electrolyte. Thermal stability study, clearly explained about the thermal behavior in DPT doped electrolyte which is a good stability electrolyte in DSSCs performances.

The PEG polymer XRD (**Figure 9**) revealed that major peaks with high intensity observed at 2θ = 19.8° (120), 23.2° (032), 27.0° (024), 27.4° (024), 31.0° (220), 36.4° (111) and 43.3° (200) for PEG/KI/I_2 and reveals that more crystalline phase. This intensity peak was matched with PEG polymer [47]. In the case of PEG/KI/I_2/DPT polymer electrolyte, showed the similar diffraction and low intensity peaks were observed, indicating that the polymer has a low crystallinity. This finding showed that the potassium iodide salt is completely miscible in the polymer matrix and reduces the crystallinity of the polymer electrolyte. When the polymer electrolyte (PEG/KI/I2/DPT) is doped with the organic compound DPT, more broadening and fewer intensities are observed, as shown in **Figure 9**. These finding suggests that the incorporation

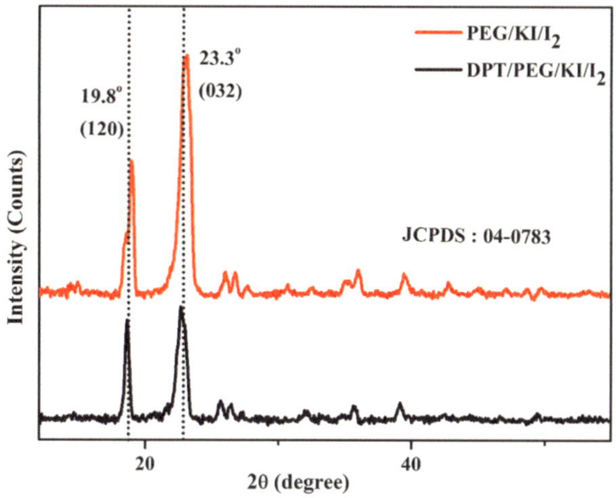

Figure 9.
XRD pattern of polymer electrolytes.

of the DPT organic compound influenced PEG crystallization and good amorphous electrolyte for DSSCs applications.

Functional groups of PEG polymer electrolyte were characterized by FTIR, ranges from 4000 to 500 cm^{-1}. **Figure 10** showed a wide absorption peak displayed at 3300–3500 cm^{-1} contributed to the O—H and N—H functional groups present in the polymer electrolyte. The C—H bending and stretching was observed at 2882–1342 cm^{-1}, O—H frequency at 1280 cm^{-1}, C—O—H stretching frequency observed at 1094 cm^{-}, respectively. All the PEG polymer functional groups exactly matched with the prepared electrolyte [48]. From the FTIR spectrum the vibration frequency intensity was changed in the DPT doped polymer electrolyte and the peak was shifted from 1450 to 1533 cm^{-1}. This statement the miscibility of the DPT organic compound doped in PEG polymer electrolyte revealed by FTIR.

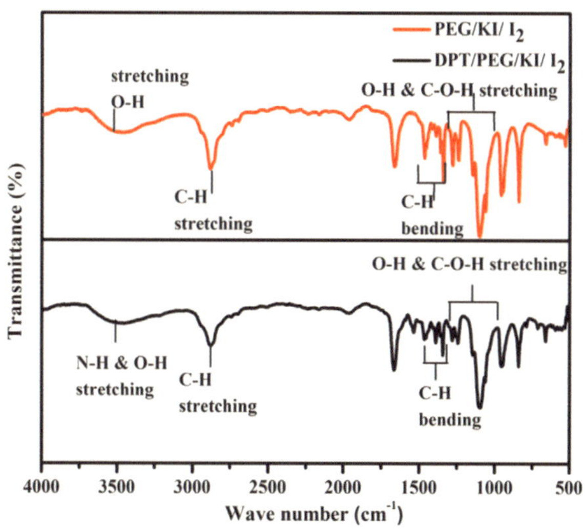

Figure 10.
FTIR spectrum of polymer electrolytes.

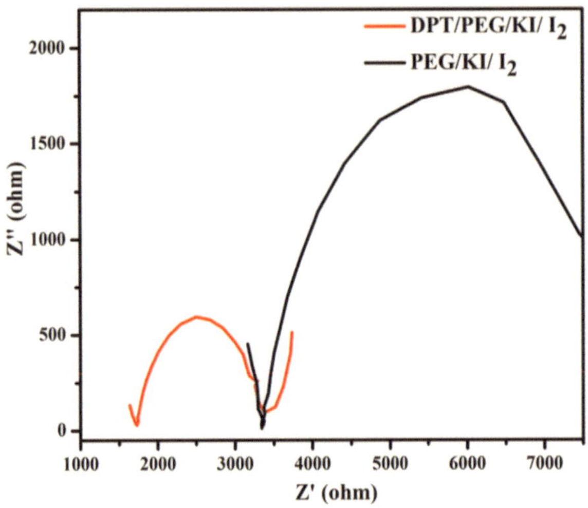

Figure 11.
Nyquist plot of polymer electrolytes.

The interfaces of GS-TiO_2/PEG/KI/I_2/DPT and GS-TiO_2/PEG/KI/I_2 were measured under dark condition during the electrochemical process. The scan range was measured at 400 kHz–300 MHz using a single sign mode of set Ewe to E of 0.800 V. In specific, the potential values were taken from the Nyquist plot which is given in **Figure 11**. Thickens of the electrolyte measured by vernier caliper. The conductivity of the electrolyte was derived from the Eq. (3).

$$\sigma = \frac{t}{R_b} A \tag{3}$$

Where, *t*—thickness of electrolyte, R_b —Bulk resistance, A—surface area [39].

The ionic conductivity values calculated for the GS-TiO_2/PEG/KI/I_2 and GS-TiO_2/PEG/KI/I_2/DPT electrolyte. The GS-TiO_2/PEG/KI/I_2 and GS-TiO_2/PEG/KI/I_2/DPT electrolyte attained conductivity at 2.9788×10^{-4} and 5.7836×10^{-4} Scm^{-1} respectively. The conductivity of DPT doped electrolyte was found to be increased. In general, the addition of KI and I_2 to the polymer electrolyte, increases conductivity. Therefore, the conductivity is accelerated by the incorporation of organic compounds. It is well known that decrease in crystallinity improves the randomness of the polymer chain, resulting in free space and improved ion mobility. The increased interaction of nitrogen present in DPT compound with iodine in the redox mediator has been recognized by the high conductivity of DPT doped in the PEG polymer electrolyte. The conductivity results supported the concept of incorporating a polymer matrix with an DPT organic compound and an I^-/I_3^- redox pair to improve the conductivity of a PEG polymer electrolyte in order to improve the efficiency and stability of DSSCs device.

3.2 I-V studies

Comparison of Com.TiO_2 and GS-TiO_2 photoelectrochemical behavior with DPT doped polymer electrolyte measured under $100 mWcm^{-2}$ sun illumination at A.M. 1.5. **Table 1** summarizes the photo electrochemical properties and showed in **Figure 12**. In

System	J_{oc} (mA/cm^2)	V_{oc} (V)	FF	Efficiency (η) %
TiO_2(p-25)/N3dye/KI/I_2/PEG/DPT/Pt	14.7	0.86	0.53	6.7
Green TiO_2/N3dye/KI/I_2/PEG/DPT/Pt	11.6	0.84	0.53	5.2

Table 1.
Photo electrochemical properties of GS-TiO_2 and Com-TiO_2 with DPT doped polymer electrolyte.

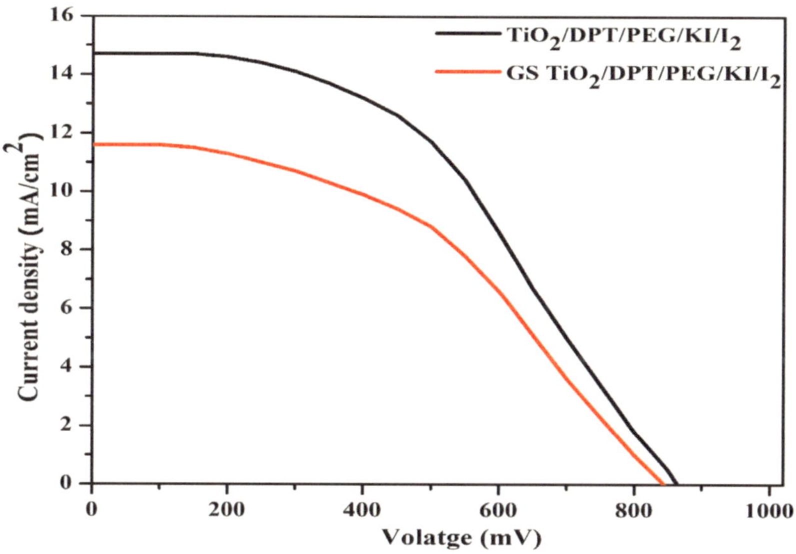

Figure 12.
I–V curves of GS-TiO_2 and Com-TiO_2 with DPT doped polymer electrolyte.

the dye-sensitized solar cell field, N3 dye coated GS-TiO_2 with PEG/KI/I_2/DPT polymer electrolyte reported a higher efficiency of 5.2%. At the same time Com.TiO_2 with PEG/KI/I_2/DPT polymer electrolyte showed the efficiency of 6.7%. The current study thus confirms the feasibility of developing DPT doped PEG polymer electrolyte with novel lab-prepared GS-TiO_2 to improve the photovoltaic properties of DSSCs. The GS-TiO_2 obtained through an environmentally friendly method has a high efficiency.

3.3 Conclusion

TiO_2 nanoparticles were effectively prepared using *A. bilimbi* fruits by green synthesis method. The GS-TiO_2 surface morphology was analyzed using SEM, UV and XRD confirmed the GS-TiO_2 is a more crystallite structure, good band gap (3.2 eV) and small particle size (15 nm). FTIR spectrum confirmed that the present of plant functional group acts as a capping agent on the GS-TiO_2 nanoparticles. The PEG/KI/I_2 electrolyte with DPT dopant improve the conductivity, stability and ionic mobility of the PEG polymer and improves the power conversion efficiency in DSSC device owing to the presence of more electron donating group in dopant DPT. The GS-TiO_2 and PEG polymer electrolyte based DSSC cell attained energy conversion efficiency of 5.2%. This way of utilizing natural TiO_2 in DSSC applications with reduced cost and chemicals requirements, eco-friendly and also the usage of PEG polymer with DPT dopant-based electrolyte for stability improvements in the DSSCs performances.

Author details

Sundaramurthy Devikala* and Johnson Maryleedarani Abisharani
Department of Chemistry, SRM Institute of Science and Technology, Chengalpattu, Tamil Nadu, India

*Address all correspondence to: devikals@srmist.edu.in

References

[1] Vipinraj S, Elsa J, Sudhakar K. Recent improvements in dye sensitized solar cells: A review. Renewable and Sustainable Energy Reviews. 2015;**52**:54-64

[2] Kumar B. Natural pigments and dye-sensitized solar cells (DSSC): A correlation. In: Engineering Practices for Agricultural Production and Water Conservation. 1st ed. Apple Academic Press; 2017. pp. 26. ISBN: 9781315365954

[3] O'Regan B, Gratzel M. High-efficiency solar cell based on dye-sensitized colloidal TiO_2 films. Nature. 1991;**353**:737-740

[4] Geetam R, Anil K, Perapong T, Bhupendra G. Natural dyes for dye sensitized solar cell: A review. Renewable and Sustainable Energy Reviews. 2017;**69**:705-718

[5] Mian EY, Kah YC. Recent advances in photo-anode for dye-sensitized solar cells: A review. International Journal of Energy Research. 2017;**41**(15):2446-2467

[6] Islam A, Chowdhury TH, Qin C, Han L, Lee J-J, Bedja IM, et al. Panchromatic absorption of dye sensitized solar cells by co-sensitization of triple organic dyes. Sustainable Energy and Fuels. 2018;**2**(1):209-214

[7] Muniandy SS, Kaus NHM, Jiang ZT, Altarawneh M, Lee HL. Green synthesis of mesoporous anatase TiO_2 nanoparticles and their photocatalytic activities. RSC Advances. 2017;**7**(76): 48083-48094

[8] Son YJ, Kang JS, Yoon J, Kim J, Jeong J, Kang J, et al. Influence of TiO_2 particle size on dye-sensitized solar cells employing an organic sensitizer and a cobalt (III/II) redox electrolyte. The Journal of Physical Chemistry C. 2018; **122**(13):7051-7060

[9] Grzmil B, Glen M, Kic B, Lubkowski K. Preparation and characterization of single-modified TiO_2 for pigmentary applications. Industrial and Engineering Chemistry Research. 2011;**50**(11):6535-6542

[10] Wang Y, Li J, Wang L, Xue T, Qi T. Preparation of rutile titanium dioxide white pigment via doping and calcination of metatitanic acid obtained by the NaOH molten salt method. Industrial and Engineering Chemistry Research. 2010;**49**(16):7693-7696

[11] Maurya IC, Singh S, Senapati S, Srivastava P, Bahadur L. Green synthesis of TiO_2 nanoparticles using Bixa orellana seed extract and its application for solar cells. Solar Energy. 2019;**194**:952-958

[12] Park NG, Van de Lagemaat J, Frank AA. Comparison of dye-sensitized rutile-and anatase-based TiO_2 solar cells. The Journal of Physical Chemistry B. 2000;**104**(38):8989-8994

[13] Frank AJ, Kopidakis N, Van De Lagemaat J. Electrons in nanostructured TiO_2 solar cells: Transport, recombination and photovoltaic properties. Coordination Chemistry Reviews. 2004;**248**(13–14):1165-1179

[14] Carp O, Huisman CL, Reller A. Photoinduced reactivity of titanium dioxide. Progress in Solid State Chemistry. 2004;**32**(1–2):33-177

[15] Gong J, Liang J, Sumathy K. Review on dye-sensitized solar cells (DSSCs): Fundamental concepts and novel materials. Renewable and Sustainable Energy Reviews. 2012;**16**(8):5848-5860

[16] Shinde PS, Bhosale CH. Properties of chemical vapour deposited nanocrystalline TiO_2 thin films and their use in dye-sensitized solar cells. Journal of Analytical and Applied Pyrolysis. 2008;**82**(1):83-88

[17] Nagaraju P, Khunphonoi R, Puttaiah SH, Suwannaruang T, Kaewbuddee C, Wantala K. Photocatalytic paraquat degradation over TiO_2 modified by hydrothermal technique in alkaline solution. Journal of Advanced Oxidation Technologies. 2017; **20**(2):1-12

[18] Wahi RK, Liu Y, Falkner JC, Colvin VL. Solvothermal synthesis and characterization of anatase TiO_2 nanocrystals with ultrahigh surface area. Journal of Colloid and Interface Science. 2006;**302**(2):530-536

[19] Bhogaita M, Yadav S, Bhanushali AU, Parsola AA, Nalini RP. Synthesis and characterization of TiO_2 thin films for DSSC prototype. Materials Today: Proceedings. 2016;**3**(6): 2052-2061

[20] Arami H, Mazloumi M, Khalifehzadeh R, Sadrnezhaad SK. Sonochemical preparation of TiO_2 nanoparticles. Materials Letters. 2007;**61** (23–24):4559-4561

[21] Falk GS, Borlaf M, López-Muñoz MJ, Fariñas JC, Neto JR, Moreno R. Microwave-assisted synthesis of TiO_2 nanoparticles: Photocatalytic activity of powders and thin films. Journal of Nanoparticle Research. 2018;**20**(2): 1-10

[22] Bandy J, Zhang Q, Cao G. Electrophoretic deposition of titanium oxide nanoparticle films for dye-sensitized solar cell applications. Materials Sciences and Applications. 2011;**2**(10):1427-1431

[23] Subhapriya S, Gomathipriya P. Green synthesis of titanium dioxide (TiO_2) nanoparticles by Trigonella foenum-graecum extract and its antimicrobial properties. Microbial Pathogenesis. 2018;**116**:215-220

[24] Rajakumar G, Rahuman AA, Priyamvada B, Khanna VG, Kumar DK, Sujin PJ. Eclipta prostrata leaf aqueous extract mediated synthesis of titanium dioxide nanoparticles. Materials Letters. 2012;**68**:115-117

[25] Nabi G, Khalid NR, Tahir MB, Rafique M, Rizwan M, Hussain S, et al. A review on novel eco-friendly green approach to synthesis TiO_2 nanoparticles using different extracts. Journal of Inorganic and Organometallic Polymers and Materials. 2018;**28**(4):1552-1564

[26] Manikandan V, Velmurugan P, Jayanthi P, Park JH, Chang WS, Park YJ, et al. Biogenic synthesis from Prunus × yedoensis leaf extract, characterization, and photocatalytic and antibacterial activity of TiO_2 nanoparticles. Research on Chemical Intermediates. 2018;**44**(4):2489-2502

[27] Alhassan AM, Ahmed QU. *Averrhoa bilimbi* Linn.: A review of its ethnomedicinal uses, phytochemistry, and pharmacology. Journal of Pharmacy & Bioallied Sciences. 2016;**8**(4):265-271

[28] Hasanuzzaman M, Ali MR, Hossain M, Kuri S, Islam MS. Evaluation of total phenolic content, free radical scavenging activity and phytochemical screening of different extracts of *Averrhoa bilimbi* (fruits). International Current Pharmaceutical Journal. 2013; **2**(4):92-96

[29] Hussain I, Singh NB, Singh A, Singh H, Singh SC. Green synthesis of nanoparticles and its potential

application. Biotechnology Letters. 2016; **38**(4):545-560

[30] Singh S, Maurya IC, Tiwari A, Srivastava P, Bahadur L. Green synthesis of TiO_2 nanoparticles using Citrus Limon juice extract as a bio-capping agent for enhanced performance of dye-sensitized solar cells. Surfaces and Interfaces. 2022; **28**:101652

[31] Wu J, Li P, Hao S, Yang H, Lan Z. A polyblend electrolyte (PVP/ PEG + KI + I_2) for dye-sensitized nanocrystalline TiO_2 solar cells. Electrochimica Acta. 2007;**52**(17): 5334-5338

[32] Ganesan S, Mathew V, Paul BJ, Maruthamuthu P, Suthanthiraraj SA. Influence of organic nitrogenous compounds phenothiazine and diphenyl amine in poly(vinylidene fluoride) blended with poly (ethylene oxide) polymer electrolyte in dye-sensitized solar cells. Electrochimica Acta. 2013; **102**:219-224

[33] Chalkias DA, Giannopoulos DI, Kollia E, Petala A, Kostopoulos V, Papanicolaou GC. Preparation of polyvinylpyrrolidone-based polymer electrolytes and their application by in-situ gelation in dye-sensitized solar cells. Electrochimica Acta. 2018;**271**:632-640

[34] Yang Y, Zhang J, Zhou C, Wu S, Xu S, Liu W, et al. Effect of lithium iodide addition on poly (ethylene oxide) —poly(vinylidene fluoride) polymer-blend electrolyte for dye-sensitized nanocrystalline solar cell. The Journal of Physical Chemistry B. 2008;**112**(21): 6594-6602

[35] Li C, Xin C, Xu L, Zhong Y, Wu W. Components control for high-voltage quasi-solid state dye-sensitized solar cells based on two-phase polymer gel electrolyte. Solar Energy. 2019;**181**: 130-136

[36] Abisharani JM, Dineshkumar R, Devikala S, Arthanareeswari M, Ganesan S. Influence of 2, 4-Diamino-6-Phenyl-1-3-5-triazine on bio synthesized TiO_2 dye-sensitized solar cell fabricated using poly(ethylene glycol) polymer electrolyte. Materials Research Express. 2020;**7**(2):025507

[37] Wu J, Lan Z, Lin J, Huang M, Huang Y, Fan L, et al. Electrolytes in dye-sensitized solar cells. Chemical Reviews. 2015;**115**(5):2136-2173

[38] Boschloo G, Häggman L, Hagfeldt A. Quantification of the effect of 4-tert-Butylpyridine addition to I^-/I_3^- redox electrolytes in dye-sensitized nanostructured TiO_2 solar cells. The Journal of Physical Chemistry B. 2006; **110**(26):13144-13150

[39] Karthika P, Ganesan S, Thomas A, Sheebarani TM, Prakash M. Influence of synthesized thiourea derivatives as a prolific additive with tris(1,10-phenanthroline) cobalt (II/III) bis/tris (hexafluorophosphate)/hydroxypropyl cellulose gel polymer electrolytes on dye-sensitized solar cells. Electrochimica Acta. 2019;**298**:237-247

[40] Bagheri O, Dehghani H, Afrooz M. Pyridine derivatives; new efficient additives in bromide/tribromide electrolyte for dye sensitized solar cells. RSC Advance. 2015;**5**(105):86191-86198

[41] Sirimanne PM, Shirata T, Soga T, Jimbo T. Charge generation in a dye-sensitized solid-state cell under different modes of illumination. Journal of Solid State Chemistry. 2002;**166**:142-147

[42] Yang L, Qin X, Jiang X, Gong M, Yin D, Zhang Y, et al. SERS investigation of ciprofloxacin drug molecules on TiO_2

nanoparticles. Physical Chemistry Chemical Physics. 2015;**17**(27): 17809-17815

[43] Isaac RS, Sakthivel G, Murthy CH. Green synthesis of gold and silver nanoparticles using *Averrhoa bilimbi* fruit extract. Journal of Nanotechnology. 2013:1-6

[44] Goutam SP, Saxena G, Singh V, Yadav AK, Bharagava RN, Thapa KB. Green synthesis of TiO_2 nanoparticles using leaf extract of *Jatropha curcas* L. for photocatalytic degradation of tannery wastewater. Chemical Engineering Journal. 2018;**336**:386-396

[45] Ong WL, Lim YF, Ong JLT, Ho GW. Room temperature sequential ionic deposition (SID) of Ag_2S nanoparticles on TiO_2 hierarchical spheres for enhanced catalytic efficiency. Journal of Materials Chemistry A. 2015;**3**(12): 6509-6516

[46] Jayaramudu T, Raghavendra GM, Varaprasad K, Reddy GVS, Reddy AB, Sudhakar K, et al. Preparation and characterization of poly(ethylene glycol) stabilized nano silver particles by a mechanochemical assisted ball mill process. Journal of Applied Polymer Science. 2016;**133**(7):1-8

[47] Bhattacharyya R, Ray SK. Removal of Congo red and methyl violet from water using nano clay filled composite hydrogels of poly acrylic acid and polyethylene glycol. Chemical Engineering Journal. 2015;**260**:269-283

[48] Shameli K, Bin Ahmad M, Jazayeri SD, Sedaghat S, Shabanzadeh P, Jahangirian H, et al. Synthesis and characterization of polyethylene glycol mediated silver nanoparticles by the green method. International Journal of Molecular Sciences. 2012;**13**(6): 6639-6650

Chapter 3

Ecological Applications of Enzymes in Plants Based Textile Dyeing

Wafa Haddar, Shahid Adeel, Mahwish Salman, Abdul Ghaffar, Mehwish Naseer, Muhammad Usama and Manel Ben Ticha

Abstract

Biotechnology has a foremost role in the textile industry by enhancing ecofriendly, cost-effective, and energy-efficient manufacturing processes. The use of enzymatic biotechnology is one of the sustainable newly developed state-of-the-art processes for textile processing. To reduce the use of toxic and hazardous chemicals, enzymes have been proposed as one of the finest promising alternatives. Many enzymes have been used widely in textile processes such as lipase, laccase, pectinase, cellulase, catalase, amylase, and protease. The enzymatic use in the textile industry is very promising because they produce top-class goods, and give way to the reduction of water, time, and energy. The increasing demand for natural dyes especially with the incorporation of enzymes makes process more sustainable and eco-friendlier to suppress the toxicity of synthetic dyes. In the first part of the chapter, particular attention has been given to the source and extraction of natural dyes. In the second part of the chapter, different enzymes and their possible roles in the textile industry have been discussed. It is expected that this chapter will provide an innovative direction to the academic researchers, the community of textile and traders as well as artisans who are working in the area of biotechnological applications for the betterment of textile processing.

Keywords: bio-processing, bio-bleaching, bio-desizing, bio-mercerization, sustainable coloration

1. Introduction

The textile industry is one of the fundamental requirements of human beings which also contributes significantly to the growing economies of many developing nations. The consumption of textile materials is expanding as a result of a growing population and increased per capita textile demand [1, 2]. However, the traditional approach of textile wet processing needs several processes by using dangerous chemicals, high salt concentrations, and a lot of resource consumption before it produces a completed fabric. This approach is criticized for its negative environmental effects due to its toxicity [3]. Implementation of enzymes in textile wet processing is guided by friendly environmental awareness. Enzymes are used in the textile sector, which appears to establish a perfect balance between commercial requirements and the creation of environmentally beneficial products [4]. Enzymatic procedures are extensively utilized in the chemical processing of textiles to reduce environmental risks and prevent the excessive use of toxic chemicals. Enzyme biotechnology is a sustainable and effective method that has been utilized in many manufacturing procedures as a favorable alternative to synthetic catalysis with benefits in terms of productivity and sustainability as well as creating top-quality textile fabrics [5]. Enzymatic catalysis is an effective instrument that can be used in the industrial setting and allows for the resolution of most environmental sustainability-related problems, particularly those involving the usage of dangerous chemicals. Enzyme utilization has the potential to significantly lessen the negative effects of industrial processes on the environment [6].

Enzymes are natural and biodegradable proteins that are frequently employed in industries to replace dangerous chemicals since they operate under compassionate circumstances and are durable, safe, and disposable. Enzymes have important roles in reducing contaminants, bio-finishing to improve esthetics, removing fabric fur from the surface, bio-bleaching of cotton to give it an excellent matte texture, and removing leftover hydrogen peroxide after bleaching [7]. Oxidoreductases and hydrolases are two major enzymes used in the textile industry. The hydrolase class comprising catalase, amylase, pectinase, lipase, and cellulose are utilized in desizing, biopolishing, bioscouring, and bleaching processes in textile industries [8]. The oxidoreductase class of enzymes comprises trans-glutaminases, which are used to modify the properties of synthetic fibers whereas laccases are used to decolorize the fabrics. Since synthetic dyes have demonstrated harmful qualities over natural dyes, researchers have drawn towards natural dyes due to their eco-friendliness [9]. The other beneficial aspects of natural dyes over artificial dyes are, the natural dyes are disposable, and the byproducts produced during the dyeing process are less harmful to the environment [10]. Moreover, natural dyes have a calming effect and provide beautiful colors and adorable shades, which make textiles appealing to consumers [11]. The majority of dye-producing plants also exhibit anti-oxidant properties, ultraviolet protection, and antimicrobial potential. Precisely, natural dyes give textiles some additional finishing qualities in addition to color [12].

2. Resources of natural dyes

Various natural sources have been used to create natural colorants, these have been categorized as plants, animals, minerals, and microorganisms [13]. Natural Dye Research and Development Project, in Turkish known as DOBAG, was launched in Turkey in 1981 in collaboration with Polytechnic University, Istanbul, and was very successful in reviving the forgotten craft of naturally colored textiles [14]. Several natural dyeing supplies have now been discovered because of research initiatives by individuals and organizations as well as information exchange at several conventions, festivals, seminars, and articles [15]. There is now a huge amount of knowledge concerning various sources of natural dyes in the literature.

2.1 Plant origin

Several natural colorants have originated from plants in history like an alkane, annatto, madder, chamomile, sappers, coreopsis, etc. Natural dye supplies include a variety of plant components such as leaves, roots, stump branches, core wood, wood shavings, bark, fruits, flowers, hulls, and husks [16]. For example, leaves of *Indigofera tinctoria* are used to make the well-known natural blue color indigo. Some plant-based dyes are used for other purposes as well, such as food coloring and traditional medicine, and as a result, there is a commercial supply chain for these dyes [17]. The commercial availability of natural dyes has expanded due to a resurgence in interest in them.

2.1.1 Blue dye

The king of natural dyes, Indigo, is the only significant natural dye in blue color which could be extracted from the *I. tinctoria* plant leaves. Since, indigo has been utilized for producing a blue hue since ancient times and is currently the most popular denim material [17]. More precisely, a pale-yellow chemical known as Indican serves as the coloring component, and it is found in the leaves of indigo plants. Interestingly, indigo plants in approximately one-acre area can produce roughly 5000 kg of leaves, which can be converted into 50 kg of pure natural indigo powder. Several plants, besides the *Indigofera* species, may be utilized to make indigo dye such as woad, a plant that naturally produces indigo in Europe. In addition, Wrightia tinctoria and Dyer's knotweed (*Polygonum tinctorium*) are two other plants that have historically been used to make indigo [18]. Unfortunately, natural indigo got declined after the production of synthetic indigo in 1987 which attained more preference over natural one.

2.1.2 Red dye

Red natural dyes can be found in a variety of plant sources. Natural red dyes called madder are made from plants of Rubia plant (Hosseinnezhad et al., 2021) "queen of natural dyes". Between 3 and 5 tonnes of roots and 150–200 kg of dye are produced per hectare by the 3-year-old plant [19]. In addition to the roots, the plant also has dye in the stems and other sections like dried root chips or stem pieces after soaking in cold water before being used to extract the dye. Being a mordant dye, it creates insoluble multiplexes with the metal ions existing on mordanted fabric to generate vivid colors. Pink and red hues are frequently produced with alum where's the alum and iron together yield purple hues. A variety of red could also be generated by mixing other mordants with the main metallic salt, alum [20]. The sappan wood often referred to as "Patang," is a tiny tree that produces a red dye often found in India, Malaysia, and Philippines. *Caesalpinia echinate*, the Brazil wood named after the word Braza which means flaming like fire due to the vivid red color of its wood, also contains the same dye.

Another red pigment-producing tree is *Morinda citrifolia* which is found in Sri Lanka and India. The 3–4 years old tree provides a good quantity of coloring matter from its bark and roots. A variety of colors, including chocolate and purple, can be produced by using different mordants [21]. Additionally, an annual herb known as safflower is believed to have come from Afghanistan and has been used to extract the dye. This herb developed an astonishing cheery red shade on silk and cotton. Dried safflower florets are repeatedly washed in acidic water to get rid of all the yellow color water-soluble material before the removal of the dye [22].

2.1.3 Yellow dyes

A well-known source of yellow dye is turmeric which is extracted from turmeric rhizomes, whether they are fresh or dried, and are used to make the color on wool, silk, and cotton. The coloring material present in turmeric is curcumin, which belongs to the diaroylmethane class [23]. To increase the fastness qualities and range of adorable shades, different mordants can be used. Saffron is also a good source of vintage yellow dye from the *Iridaceae* family that is made up of the desiccated stigmas of the *Crocus sativus* plant. By boiling flower stigmas in water, the dye is released and produces a vivid yellow hue on cotton, silk, and wool [24]. *Berberis aristata* commonly known as the barberry plant is also a well-reputed source of yellow dye. The roots, bark, and stems of the barberry plant are used to extract the yellow dye and can be used directly to color silk and wool with average lightfastness and good washing fastness properties [25].

Another source, pomegranate (*Punica granatum*) fruit rinds, which are high in tannin, are used for mordanting. It is also used in conjunction with turmeric to increase the dyed fabrics' light resistance. Myrobolan (*Terminalia chebula*) fruits also have high tannin content. The dried fruit also contains a natural colorant which develops a vivid yellow color for all textile fabrics. Moreover, myrobolan can also be employed as a natural mordant for natural dyes fixing on textile fabrics.

Marigold (*Tagetus spp.*) is also frequently employed to create garlands and floral accents due to its vivid yellow flowers. It comes in a variety of colors, such as yellow, golden yellow, orange, and others [26]. The primary yellow coloring agents are Quercetagetol, a flavonol together with two of its glycosides, and lutein which exerts good fastness characteristics of wool and silk. This dye can be used to quickly create colors on cotton when combined with mordants. Besides, the flame of the forest tree named *Butea monosperma* color all natural fibers. By using correct mordants, the dye extracted from the bright orange flowers of this tree develops vivid yellow, brown, and orange colors. *Mallotus sphillipensis's* dried fruit capsules, known as Kamala, produce a reddish-orange powder. A vibrant yellow-orange and yellow-golden hue developed onto silk and wool [27]. Similarly, the outside layer of onions, *Allium cepa*, which is typically discarded like trash, be able to be utilized to extract natural yellow colorant. The chemical makeup of the dye is a flavonoid, and it gives wool and silk vibrant colors. A suitable mordant can be used to dye cotton with average washing and light resistance properties.

2.1.4 Black and Brown dye

Oak galls are utilized for mordanting because they are high in tannin and also be employed to achieve a brown hue. Catechu or cutch, which is made from the heartwood of *Acacia catechu* employed to dye cotton, wool, and silk [28]. It has a lot of tannins as well and can be dyed black using iron mordant. By iron mordanting, many yellow and red dyes can also be turned black. Likewise, the heartwood of the Haematoxylon campechianum tree, which is located in the West Indies and Mexico, was used to extract the famous logwood black hue, which is quite sharp and has excellent fastness capabilities [29].

Some other research reports the valorization of some natural wastes such as olive wastewater by its use as a possible dye bath for dyeing textile fibers. During olive oil extraction, dark brown to black effluent was produced which was characterized by a high organic load including polyphenols and tannins. Darker brown shades were developed with generally good fastness which demonstrates that protein fibers, cotton, and other synthetic fibers possess a high affinity to this aqueous extract.

3. Natural dyes production and extraction techniques

Natural colors are typically derived from diverse plant parts, distinct from synthetic dyes, which are created from chemical predecessors. These dye-bearing materials typically only have a 0.5–5% dye content and these plant ingredients cannot be used directly for the dyeing process [30]. Additionally, a lot of plant materials, including flowers and fruits, are seasonal and contain a lot of water, making it impossible to store them in their natural state. Therefore, these are put through some processing procedures to make them appropriate for dyeing in textile industry needs and to make them accessible all around the year [31]. To lower their contents of water to about 10–15% or less, collected constituents of plants are desiccated at first, either in the shadow or at a low temperature of 40–50°C in a hot air dryer. To minimize particle size and improve dye extraction, the dried material is subsequently ground into a powder [11]. In most situations, these powdered and dried components can be kept for at least a year in sealed bags or containers and utilized for dying whenever necessary. To create pure dye powders, the dye must first be extracted from materials containing dye. Due to the use of several types of machinery and higher energy consumption throughout various processing activities, these refined versions are expensive [32]. Additionally, because dye extraction happens simultaneously with dyeing, its effectiveness is lower when compared to using powdered crude dye-bearing material [33].

Since the amount of coloring matter or dye present in natural dye-bearing materials is relatively low, along with other plant and animal compounds like water-insoluble fibers, carbohydrates, protein, chlorophyll, and tannins, among others, extraction is a crucial step both in the production of purified natural dyes as well as in the processing of raw dye-bearing materials [34]. Before using an extraction procedure, the type and solubility characteristics of the coloring components must be studied [35]. Many conventional and non-conventional techniques for extracting colored ingredients are described in the following.

3.1 Conventional extraction methods

3.1.1 Aqueous extraction

Plants and other materials were previously utilized to extract colors using aqueous extraction. To increase the effectiveness of the extraction process, the material which is comprising dyes is at first fragmented into tiny bits or powdered before being sieved [36]. It is then immersed in water for a long period typically overnight in earthen, wooden, or metal vessels (ideally copper or stainless steel) to release the cell structure [37]. The dye solution is then extracted and filtered to eliminate any remaining non-dye plant material. To get rid of the non-dye parts, the boiling and filtering operation is repeated, and centrifuges are typically used to separate leftover material. The elimination of tiny plant materials and improved solubility of the purified natural dye can both be achieved by using trickling filters [29].

The extract generated by this procedure may be employed to the constituents of the textile with ease because many dyeing processes are conducted in aqueous solutions. This extraction method has several drawbacks, including a lengthy extraction period, a significant amount of water needed, the usage of elevated temperatures, and a negligible yield of dye because merely the water-soluble color constituents are removed, even though numerous dyes comprise little water solubility [38]. As well as the dye, other water-soluble materials are also extracted, these materials may need to be eliminated if the extract is to be condensed and turned into a fine powder. Boiling temperature reduces the yield of heat-sensitive dye compounds therefore a low temperature should be optimized for the extraction in these circumstances [39].

3.1.2 Acid and alkali extraction process

Many colors exist as glycosides, they can be removed using diluted acidic or alkaline solutions. Greater extraction and elevated yield of coloring constituents occur from the hydrolysis of glycosides being facilitated by the addition of acid or alkali [40]. Tesu (*B. monosperma*) flower petals are utilized to extract the dye by an acid hydrolysis procedure. To avoid oxidative degradation, some flavone dyes are extracted using acidified water [41].

Alkali and alkaline extraction are appropriate for dyes with phenolic groups as it increases the yield of the color. This method is also used to extract rose-red colorant from petals safflower. One drawback of this procedure is that few dyeing components might be degraded in alkaline environments because some natural dyes are pH-sensitive [42]. The colorants which exist naturally are typically a combination of many biochemical components, altering the pH of the extraction medium by complementing alkali or acid able to cause the extraction of various colorant components, which can result in a range of color outcomes and colorfastness characteristics [43]. To determine the ideal optimization for dye extraction, numerous scholars have investigated natural dye extraction under numerous conditions of pH and equated the fastness and shade attributes of tinted fabrics [44].

3.1.3 Solvent extraction

Depending on their nature, natural coloring substances can also be extracted by utilizing natural solvents like chloroform, methanol, petroleum ether, acetone, ethanol, or mixtures of solvents like ethanol and methanol, water, alcohol, etc. [45]. Both water-soluble and water-insoluble materials can be extracted from plant resources using the water/alcohol extraction method. As a result of the ability to extract a greater variety of chemicals and coloring ingredients than the aqueous approach [46]. Alcoholic solvents may also be used with an acid or alkali to aid in the hydrolysis of glycosides and the release of coloring components. The ability to readily remove and reuse solvents makes it simpler to purify extracted colors. Because extraction is done at a lower temperature, there is less danger of deterioration [47]. The method's drawbacks include the greenhouse effect of the poisonous leftover solvents and requiring an aqueous solution for the dyeing process because the extracted substance is not easily soluble in water. Problems can also result from the co-extraction of compounds like waxy polymers and chlorophylls [41].

3.2 Non-conventional extraction methods

3.2.1 Ultrasonic and microwave extraction

These are ultrasound and microwave-assisted extraction techniques, in which the use of ultrasound or microwaves improves extraction efficiency and reduces the amount of solvent needed as well as the extraction time [48]. Ultrasound induces the formation of tiny bubbles or cavitation in the liquid when the plant components are preserved with water or some additional solvent. This results in the cavity collapsing or the bubbles bursting, which raises the temperature and pressure [49]. The extraction efficiency is quickly increased when extremely high temperatures and pressures are created. Many studies have lately reported using this extraction approach as the quest for novel dye sources and efforts to improve dye extraction continue [50].

In microwave extraction, the natural sources are processed in the existence of microwave energy sources with the least amount of solvent possible. The procedures are accelerated by the microwave, allowing for faster and more effective extraction [51]. The decrease in temperature of extraction, utilization of solvent, and duration show consequences in less energy utilization. Both extractions ultrasound and microwave may be regarded as green methods [52].

3.2.2 Fermentation

Fermentation accelerates the process of extraction by using the enzymes which are generated by microbes found in the environment or natural resources. The most typical instance of this kind of extraction is indigo extraction in which newly collected indigo twigs and leaves are immersed in warm water (about 32°C) [38]. As fermentation progresses, the indimulsin enzyme, which is also present in the leaves, converts the colorful indigo-containing glucoside indican into glucose and indoxyl. In about 10–15 hours, fermentation is finished, and the indoxyl-containing yellow fluid is formally transferred to whipping tanks where indoxyl is became oxidized through the air and turned into the blue color, insoluble indigotin that sinks to the bottom [53]. It is collected, cleaned, and then pressed to remove the extra water. Other colorants, such as annatto, can also be extracted using this method. Except for not requiring high temperatures, the fermentation process is comparable to aqueous extraction [28]. The bacteria naturally break down the chemicals that bind coloring materials. The drawbacks of this method include a lengthy extraction process, the requirement to extract pigments right away after harvesting, a bad odor brought on by microbial activity, and others [54].

3.2.3 Supercritical fluid extraction

Supercritical fluid extraction is a developing field in the purification and extraction of natural products. Above its critical temperature and pressure, a gas behaves as a supercritical fluid. A fluid like this has physical characteristics that fall midway between a liquid and a gas [55]. They have substantially lower surface tension than liquids, which allows them to spread out along a surface more quickly [56]. As a result of their low viscosity and excellent diffusivity, they interact with the substrate more effectively. The ability to dissolve the matter in every solvent is increased at elevated pressures and temperatures, and these circumstances are required to sustain a gas in supercritical conditions.

Carbon dioxide (CO_2) supercritical fluid extraction is a viable substitute for solvent extraction since it is inexpensive, simple to use, non-toxic, and residue-free. CO_2 supercritical extractions normally take place between 32 and 49°C between 1070 and 3500 psi of pressure [57]. The procedure has acquired popularity in the extraction of purely natural ingredients for culinary and medicinal uses because the extract is off-loaded from leftover traces of solvent, and heavy-weight metals, and is bright colored because of the lack of polar polymerizing chemicals. The method's drawbacks include expensive equipment costs and polar chemical extraction [58].

3.2.4 Enzymatic extraction

Commercially accessible enzymes like pectinase, cellulase, and amylase have been utilized by certain scholars to loosen the nearby component, allowing the extraction of dye molecules under more benign situations [59]. This is because plant tissues comprise cellulose, starches, and pectins as fixing components. This method might be useful for getting dye out of tough plant components like bark, roots, and the like [60].

4. Microbial-origin enzymes in textile industry

The use of enzymes in the textile industry is an example of white/industrial biotechnology, which allows the development of environmentally friendly technologies in fiber processing and strategies to improve the final product quality [61]. The enzymes in the textile industry is an example of white/industrial biotechnology, which allows the development of environmentally friendly technologies in fiber processing and strategies to improve the final product quality. The enzymes utilization is an illustration of white/modern biotechnology that improve the final product quality and permits the expansion of technologies that are approachable to the environment [62].

Various microbial enzymes are utilized in the clothing industry at various stages on behalf of finishing and waste degradation purposes. Because of its less harmful technology and extremely low waste production, the enzyme in textiles is responsible for a big profit worldwide [63]. Amylases are frequently utilized for the desizing process in the preliminary finishing area, and cellulases are frequently used for softening, bio-stoning, and lowering the pilling tendency for cotton items in the finishing area [64]. Microorganisms are employed for enzyme manufacturing because they have a rapid capacity to adapt to any condition and can create a wide variety of enzymes. *Bacillus amyloliquefaciens*-derived α-amylase worked at pH 6.5 and 60°C for one hour with 100% desizing efficiency [65]. Amylase extracted from *Aspergillus niger* and *Aspergillus flavus* has increased desizing efficiency (*A. niger* 96%, *A. flavus* 90%), and its absorbency and controllable impurities have significantly improved [66]. Chitosan reduces the amount of enzyme needed by two-thirds while improving the desizing effect when combined with mesophilic amylase is thermally stable at high-temperature desizing [67]. A thermostable cellulase extracted from *Talaromyces emersonii* is used to treat jute-based fabrics and show greater brightness, handling, and enduring softness. Due to flavonoid oxidation, laccase from *Trametes hirsute* with mediator improves cotton's whiteness [10]. To provide whiteness, the complex enzymes Laccase and Peroxidase effectively break down and eliminate lignin from flax fabrics [68]. Another key benefit of adopting microbial enzymes is that the microorganisms are easily biotechnologically altered to produce more enzymes.

The two primary classes of enzymes utilized in the pre-treatment of cotton are oxidoreductase and hydrolase. Pectinase, a crucial enzyme extracted from the widespread bacterium *Bacillus subtilis*, is utilized to increase the scouring impact of cotton fibers whereas for the processing of cotton, catalase from *Aspergillus flavus* is utilized. *Bacillus* appears to be a relatively widespread microbe that produces a variety of enzymes that are extremely useful in the finishing process for textiles. Applications for enzymes include the fading of both denim and non-denim, bio-scouring, bio-polishing, finishing wool, removing peroxide, decolorizing dyestuff, etc. [64] (**Table 1**).

Enzyme	Source	Use in textile	References
Cellulase	*Bacillus sp., Streptomyces albaduncus, Aspergillus oryzae, Trichoderma reesei, Chaetomium globosum, Hypocrea jecorina, Trichoderma viride* G, *Aspergillus nidulans* AJ SU04	Bio-stoning, bio-polishing, and softening of denim	[69]
Catalase	*Micrococcus luteus, Bacillus sp., Bacillus cereus, Flavobacterium sp., Bacillus pumilus.*	Used after bleaching for cotton processing, Biostone washing, Bio polishing of cotton fabrics	[70]
Amylase	*Bacillus sp., S. albaduncus, S. albaduncus, Bacillus sp.* SI-136, *Bacillus sp.* SI-136, *B. cereus, Aspergillus niger, Aspergillus flavus, Aspergillus niger* SH-2, *Penenzim HSE, Bacillus subtilis* MTCC 121, *Aspergillus tamari, Aspergillus tamari.*	Remove starchy layer, De-sizing of cotton fabrics	[71]
Protease	*Bacillus licheniformis, Arthrobacter, Streptomyces, Flavobacterium sp., Bacillus sp.*	Prevent decolonization of denim, antifelting finishing treatment on wool and silk fabrics	[72]
Pectinase	*B. subtilis, Paecilomyces variotii, B. pumilus* AJK, *Streptomyces griseus, Candida.*	Hydrolysis of pection in cotton fiber preparation, scouring of cotton	[73]
Lipase	*Aspergillus niger, Candida cylindracea, Candida rugose, Streptomyces acrimycini* NGP 1, *S. albogriseolus* NGP 2, *S. variabilis* NGP, *B. licheniformis, Pseudomonas fluorescens, Bacillus sonorensis, Thermomyces lanoginosus.*	Modification of Polyester fabrics, Bio-scouring of cotton fabrics, Surface modification of Polyethylenetere-phthalate fibers	[74]

Table 1.
Microorganism-based enzymes and their role in textile.

5. Enzymes in textile

Due to its nontoxic and eco-friendly qualities, enzyme utilization in the textile sector is increasingly getting attention around the world. They also have the advantage of being able to work on specified substrates [75]. Some of the main enzymes used in processing textiles are discussed in the following.

5.1 Amylases

The most widely utilized enzyme in the clothing industry is amylase which works by breaking the starch molecules to make various compounds, for instance, dextrins and ever-smaller polymers made of glucose units [76]. The two types of starch-hydrolyzing enzymes, α-amylases, and β-amylases are categorized based on the kind of sugars they create. Various fungi, yeasts, and bacteria produce amylases, however the enzymes most frequently employed in the industry are from filamentous fungi and bacteria. Most fungi and bacteria's α-amylases are constant throughout a varied pH scale, from 4 to 11 [77]. Typically, it serves as a desizing agent by eliminating the sizing components (which are starch and its derivatives) without harming the fabric. The α-amylases enzyme hydrolyzes the $\alpha\text{-}1 \rightarrow 4$ glycosidic linkages more quickly than the β-amylases, that's why it is also known as alpha-enzyme [78]. Amylases are occasionally added including other enzymes (pectinases and cellulases) during processes like bio-scouring to lower the operating costs.

Traditionally, size materials have been put into the warp yarns to ensure seamless weaving. The majority of the ingredients used to scale the warp yarns are starch-based. Before bleaching and coloring, the sizing protecting layers from the yarns must be eliminated [79]. De-sizing refers to the procedure of eradicating the size components from the warp yarns. Traditionally, desizing is accomplished by applying chemicals (acid, alkali, oxidizing agents) to the woven cloth at higher temperatures. Furthermore, the starch cannot be adequately removed using these traditional methods, which results in inconsistent coloring. The chemicals used in the traditional procedure degrade cotton fabric as well, which results in cotton cloth losing its natural feel. After the procedure, the leftover chemicals are released into the environment, which seriously pollutes the ecosystem [80]. Enzymatic desizing is regarded as the industry's first commercialized use of biotechnology in the textile industry. The enzyme amylase is applied to the woven cloth to break down the starch, which then cleans the warp thread. The α-amylase is one of these three amylases that are best suited for the hydrolysis of starch [81]. These enzymes create dextrin, which is easily removed by water, by breaking the connection between both the glucose molecules in the starch polymer. The use of chemicals will be reduced due to the lower treatment temperature and the gentler pH condition. Although a wide range of yeasts, fungi, and bacteria are employed to manufacture α-amylases [79]. According to a study, using ultrasonic waves increases the efficiency of amylase enzymes during the de-sizing process. Amylase can be used with hydrogen peroxide or cellulase to increase the pretreated fabric absorbency, dye uptake, and rigidity. Novozyme invented a commercial alkali stable enzyme that can be employed in a wide pH (6–10) and temperature (30–90°C) range [82]. It has also been found that immobilized amylase, is less susceptible to pH, chemicals, and temperature [83]. The immobilized amylase has potential industrial applications using the magnetic cross-linked enzyme aggregation technique. For prolonged usage, this immobilized enzyme is simply detached under a magnetic field [84].

5.2 Pectinases

The middle lamella of plant cell walls contains complex polysaccharides like pectin and other compounds. The complex group of enzymes known as pectinases is responsible for the breakdown of these compounds. In nature, saprophytes and plant pathogens (bacteria and fungi) largely create them for the breakdown of plant cell walls. Pectin degrading enzymes fall into three categories: polygalacturonases (PGs), pectin esterases (PEs), and polygalacturonate lyases (PGLs) [85]. In addition to bacteria and fungi, pectin esterases are primarily produced by plants like tomatoes, citrus fruits, and bananas. Most microbial and plant PEs have a molecular weight of 3050 kDa where the ideal pH for their activity ranges from 4.0 to 7.0. Polygalacturonases are a class of enzymes that hydrolyze α-1,4 glycosidic bonds in pectin by utilizing mutually endo- and exo-splitting methods [77]. These enzymes are frequently found with molecular weights of 3080 kDa and work at an ideal temperature between 30 and 50°C, and their ideal pH range from 2.5 to 6. Although PGL from *Bacillus licheniformis* remained ideal in the pH range between 8.0 and 7.0. Although PGL from thermophiles has an optimal range between 50 to 75°C, the optimal range for PGL activity is often between 30 and 40°C [76].

The cuticle and primary wall of a raw cotton fiber include natural contaminants or noncellulosic substances (oil, proteins, waxes, pectin, lipids). To create absorbency cotton fibers, non-cellulosic elements are eliminated by the process of scouring which is boiling the gray cotton fabric in a 3% alkaline solution [86]. This conventional method has several drawbacks, including using large amounts of chemical compounds, necessitating rapid rise after scouring to make the fabric neutral, reducing the fabric weight by 6%, weakening cotton fabrics due to the formation of oxycellulose in the presence of oxygen, and usually requires more energy due to boiling temperatures [87]. The wastewater that results also has a high salt content, a great chemical oxygen demand (COD), and a high biological oxygen demand (BOD). Enzymatic scouring has been looked into in this situation as a potential replacement for conventional alkaline-based scouring [88]. In comparison to the conventional method, bioscouring has many benefits, including less water usage and energy-saving treatment temperatures. Since enzymes do not destroy cotton, unlike conventional scouring methods, there is no loss of weight or strength. Pectinases are frequently employed for scrubbing cotton as it works by breaking down the pectin in cotton's main cell wall [89]. The noncellulosic material is kept linked to the cellulose by pectin, as a result, the other contaminants can be easily separated if the pectin is eliminated. Polygalacturonases, pectin esterases, and polygalacturonate lyases are the three groups of pectinases [90]. The optimum temperature and pH range of 30–50°C and 05–08 has been reported for pectin esterases to catalyze the hydrolysis of pectin methyl esters and convert them into pectin acid. Glycosidic links in pectin are hydrolyzed by polygalacturonases which optimally work in a temperature range of 40–60°C and acidic pH 03–07 [91]. Pectinases and cellulases are usually used in combination for bio-scouring, pectinases break down pectin in this conjunction whereas cellulases demolish the cuticle by rupturing the main wall. The outcomes demonstrated that the immobilized enzymes were as effective as the combination of pectinase and cellulase and this approach is more efficient than aqueous bio-scouring [92]. High temperatures and alkaline environments are required to increase the activity and stability of pectinase. *Bacillus pumilus* BK2 has been reported as a novel source of pectate lyase, with optimal activity at pH 8.5 and a temperature of 70°C to evaluate the bio-scouring of cotton fabric [87]. To well comprehend the procedure of bioscouring and its impacts on textile materials, it is crucial to characterize the biochemical and physical surface modifications of clothes following bio-scouring and to identify effective methodologies for surface characterization [4].

5.3 Laccases

Multicopper enzymes called laccases catalyze the oxidation of a variety of phenolic and non-phenolic substances by reducing molecular oxygen by four electrons to produce water. Laccases are most common in fungi, though they have been discovered in plants, insects, and bacteria. More than 60 fungi species have been shown to produce laccase having a molecular weight of 6070 kDa, with an ideal pH range of acidic, and an ideal temperature range of 50 to 70°C [73]. Laccase from *Ganoderma lucidum* is one of the enzymes with optimal temperatures below 35°C, it works at an optimum temperature of 25°C. Laccases have several different roles in the treatment of textiles, including finishing fabric, bleaching, scouring, and dying wool, as well as playing a part in water treatment and dye synthesis. The laccases are extensively researched for denim bleaching to substitute stone washing because they can destroy indigo [93]. Research on laccases for the decolorization of textile effluents is extensively used as an eco-friendly method for treating dye wastewater because they have the potential to degrade a wide range of chemical compounds including synthetic dyes [94].

Conventionally, cotton is bleached by discoloring natural pigments, which gives white appearance to cotton fabric. Cotton has natural pigments, primarily flavonoids, which give its inherent grayness. The procedure of bleaching is used to take out the textile's natural colorants [95]. Hydrogen peroxide is a conventional market bleaching agent that works at pH 11–13 and temperatures up to 130°C to bleach materials. The traditional bleaching technique has significant drawbacks due to high temperatures and alkaline pH which seriously harm the fabric [96]. The bleaching compounds can also reduce cotton's degree of polymerization, which create overall damage to the fabric. Additionally, more water is needed after bleaching to neutralize the fabric and eliminate extra hydrogen peroxide. The use of enzymes can solve the issues with the conventional procedure [97]. Cotton is bleached by laccases which oxidized the flavonoids present in the fabric. These enzymes are utilized at temperatures between 60 and 80°C with an acidic pH, where the use of ultrasonic energy could enhance laccases' bleaching effect [98]. Besides that, glucose oxidases can generate hydrogen peroxide and gluconic acid in an aqueous solution by oxidizing glucose at acidic pH to neutral and lower temperature. It is feasible to reuse the desizing bath as a source of glucose, this biochemical method offers combined desizing and bleaching [99]. The excess hydrogen peroxide is traditionally eliminated after bleaching cotton cloth by using a reductant or by rinsing it with water. Besides this traditional method can be replaced by catalase which converts hydrogen peroxide into water and oxygen [100]. Previous studies reported a laccase-assisted wool dyeing technique that uses low temperatures and no dying auxiliaries and prevents the excessive use of water and energy [101]. In a more recent study, laccases were used to catalyze an enzyme process that used natural flavonoids to color cotton [102].

5.4 Cellulases

Cellulases are hydrolytic enzymes that speed up the process of cellulose breaking down into smaller oligosaccharides and then glucose. Cellulase activity signifies the combined action of at least three different types of cellulases in a multicomponent enzyme system [103]. Combinations of all three varieties of enzymes have better activity than the total activities of every enzyme alone because cellobiohydrolases work synergistically with endoglucanases and each other. Exo-cellulases generate cellobiose and soluble oligosaccharides, which are then transformed into glucose by β-4-glucosidase [104]. Numerous fungus cellulases are modular proteins made up of a connecting linker, a carbohydrate-binding domain (CBD), and a catalytic domain (CD). CBD serves as a mediator for the enzyme's attachment to the substrate of insoluble cellulose [105]. Temperatures between 30 to 60°C are the active range for cellulases, but they are categorized as acid-stable (pH 4.55), neutral (pH 6.67), or alkali stable depending on how sensitive to pH (pH 9.10). Cellulases were first time utilized in the textile processing industry for the finishing of denim in the late 1980s. Cellulases are currently utilized to treat cotton and other cellulose-based fabrics in addition to bio stoning [86].

Many clothes are given a washing action to provide them a somewhat worn appearance, such as stone-washing of denim jeans, which causes the blue denim to fade due to the pumice stones. This ancient method has some drawbacks, including difficulties in removing pumice residue from denim clothing, harm to the machinery and clothing, dust in the washing equipment, and ecological damage [106]. The use of pumice stone has decreased or has been entirely removed in the industry of textile due to the introduction of cellulase enzymes. In the business of textile, indigo dyes are typically used to color fabric, whereas cellulase enables the surface of the fabric to hydrolyze, which eliminates some of the indigo dyes from the surface of the fabric and gives it an aging and faded appearance [107]. Part of the indigo is removed from the fiber's surface through partial hydrolysis, resulting in bright patches. The current research focuses on the prevention or improvement of back staining, which is the redeposition of liberated indigo onto the clothing. According to reports, back staining issues can be reduced by utilizing neutral and endo cellulases at neutral pH. Purified and characterized 20 and 50 kDa endoglucanase, as well as a 50 kDa cellobiohydrolase, were previously reported for their use in textile processing at neutral pH [75]. The 20 kDa endoglucanase performed well during biostoning, and it was feasible to reduce the amount of back staining by joining the 50 kDa cellobiohydrolase or 50 kDa endoglucanase with the 20 kDa endoglucanase [108]. This problem is also solved by using enzyme-based anti-black staining agents which are composed of lipases and proteases. An important area of research has been the optimization of biofinishing procedures to collect the enzyme and reuse it. The preventive measures that have to be adopted for the usage of cellulases are that it should operate with deliberate kinetics so that no impairment appears to the internal composition of fiber and the procedure should be confined to the hydrolysis of only unfastened surface fibrils [103]. This issue can be solved by selecting the right immobilized enzyme, adjusting the concentration and incubation duration, using liquids with varying viscosities, making foam ingredients, and using hydrophobic agents to impregnate clothing.

5.5 Serine proteases: Subtilisins

Alkaline serine proteases belonging to the subtilisin family are typically extracted from numerous *Bacillus* species. They create an intermediate acyl-enzyme that accelerates the hydrolysis of peptide and ester linkages. Subtilisins are made as preproproteins precursors, where's the active site of the enzyme made of a catalytic triad of aspartate, serine, and histidine [2]. Majorly subtilisins have a molecular weight between 15 and 30 kDa, however, there is a small number of anticipations like *B. subtilis* subtilisin having a molecular weight of 90 kDa. Alkaline proteases have an optimum temperature range of 50 to 70°C, where they function at their best, but they are relatively stable at higher temperatures [109]. Thermos ability of enzyme is increased when one or more calcium binding sites are present. Subtilisins can be effectively inhibited by diisopropyl-fluorophosphate (DFP) and phenyl methyl sulphonyl fluoride (PMSF). Therefore, most subtilisin protein engineering has concentrated on improving thermostability, substrate selectivity, and oxidation resistance [33].

Raw wool is hydrophobic because of the epicuticle surface membranes comprising fatty acids and hydrophobic contaminants such as grease and wax. Alkaline scouring with sodium carbonate and preparation with potassium permanganate, sodium sulfite, or hydrogen peroxide is common methods to remove these contaminants [72]. When wet processed, wool cloth has a propensity to feel and shrink, and several chemical techniques can be used to control the way wool shrinks. The chlorine-Hercosett technique, which has been used for more than 30 years, is the very effective industrial shrink-resistant method now existing [110]. There are some significant drawbacks to this method, despite its advantages (noble anti-felt influence, little destruction, and little loss of weight), including its restricted endurance, inadequate treating features, yellowing of the fibers, challenges with coloring, and influence of the environment from the liberation of absorbable organic halogens [111]. Previous studies have recommended treating wool using safe chemical techniques, like low-temperature plasma. Plasma treatment, which is a dry method, uses electric gas discharges to treat wool fabric, since no chemicals are used in the process and only the surface characteristics of the wool are altered, it is considered as being environmentally [112]. However, the commercialization of a plasma treatment technique is limited by costs, compatibility, capacity, and the shrink-resist qualities acquired do not impart a machine-washable finish, which is one of the primary goals. A natural polymer, like chitosan, may then be used to enhance the wool's anti-felting or shrink-resistance qualities [113].

Proteases of the subtilisin form have lately been investigated as a substitute for chemically pre-treating wool, primarily for environmental concerns. According to much research, pretreating wool fabric with proteases enhanced its anti-shrinkage qualities, eliminate contaminants, and raised dyeing affinity. The enzyme can, however, enter the fabric cortex because of its small size, which leads to the degradation of the internal structure of the wool fabric [114]. According to several studies, increasing the size of the enzyme through chemical cross-linking with glutaraldehyde or by attaching synthetic polymers like polyethylene glycol might minimize the enzyme penetration, which in turn lowers potency and weight loss [115]. By increasing the cuticle's sensitivity to proteolytic decomposition, hydrogen peroxide pretreatment of wool fabric at an alkaline pH in the existence of elevated salt concentrations also focuses the activity of enzymes on the wool's outer surface [100]. As an alternative to the current proteases, the search for novel protease-producing microorganisms with great cuticle specificity is being researched.

5.6 Nitrilases and nitrile hydratases

About 40 years ago, the nitrile-hydrolyzing enzyme nitrilase was originally identified. After studying the composition and amino acid sequence of nitrilase, 13 divisions make up the nitrilase superfamily. In contrast to the eight or more branches that appear to have evident amidase or amide concentration activities, associates of only a single division are identified to exhibit the actual activity of nitrilase [116]. A small number of fungal species and 3 out of the 21 plant families have this enzyme function, however, it is more frequently found in bacteria. It is known that certain taxa, including *Pseudomonas, Klebsiella, Nocardia*, and *Rhodococcus*, use nitriles as their only carbon and nitrogen sources [117]. Different bacteria and fungi that can hydrolyze nitriles have been discovered, mostly because of the biotechnological potential of nitrilases. The majority of the isolated nitrilases were composed of one polypeptide with a molecular weight of 3045 kDa, which under certain conditions would assemble to create the active holoenzyme [118]. The enzyme appears to exist most frequently as a big aggregate with 626 subunits, although increased levels of solvents which are organic in nature, temperature, pH, salt, or even the enzyme itself can cause subunit interaction and subsequently stimulation, whereas the majority of enzymes exhibit substrate-dependent activation [119]. The primary enzyme in the enzymatic process for converting nitriles to amides, which are then transformed by amidases to the appropriate acid, is nitrile hydratase (NHase). Several microorganisms with NHase activity have been isolated, and the enzymes have been refined.

Excellent characteristics of polyacrylonitrile fiber (PAN), including their great biochemical conflict, superior flexibility, and ordinary-looking esthetic qualities, have led to a rise in their use; at the moment, they account for 10% of the synthetic fiber market worldwide [3]. However, PAN textiles' hydrophobic characteristic imposes unwanted qualities, making the dyeing and finishing procedure challenging. The surface chemical hydrolysis of PAN fabrics typically causes an irreversible yellowing of the fabric [120]. Selective enzymatic hydrolysis of PAN could therefore be a fascinating alternative, just like with other synthetic fabrics. Different sources (*Rhodococcus, Rhodochrous,* and *Agrobacterium tumefaciens*) of nitrile hydratase and amidase affected the surface of PAN [121]. The fabric's hydrophilicity and dye absorption improved after enzymatic treatment. In previous studies, PAN was treated with nitrile hydratases extracted from *Brevibacterium imperiale, Corynebacterium nitrilophilus*, and *Arthrobacter sp*. showed an increased number of amide groups on the PAN outward, increasing its hydrophilicity and dye ability [122]. In different investigations, it was discovered that the *Micrococcus luteus* strain BST20 produces membrane-bounded nitrile hydrolysing enzymes, which were shown to hydrolyze nitrile groups on the PAN surface by determining the NH_3 release from PAN powder and the depth of shade of enzyme-treated fabric following dyeing with a basic dye [123]. Matamá et al. [57] demonstrated the biomodification of acrylic fabric by utilizing a nitrilase rather than nitrile hydratases/amidases. The catalytic efficiency was increased by adding 4% N-N-dimethylacetamide and 1 M sorbitol to the treatment solution. Previous findings show that the enzymatic action of PAN will improve the characteristics of treated fabric while also saving energy and reducing pollutants, even though there is not an industrial application for it yet.

5.7 Lipase/esterases

Acyl-hydrolase enzymes sometimes referred to as lipases found in a vast variety of animals, plants, and microorganisms, accelerate the breakdown of triacyl glycerol into fatty acids and glycerol. These enzymes exhibit excellent regio- andstereo specificity a broad substrate tolerance, and other properties that make them desirable biocatalysts for the synthesis of fine chemicals and the creation of optically pure molecules [124]. They are often quite stable, do not require cofactors, and even function in organic solvents [96]. For lipases and esterases, the mechanism for ester hydrolysis or production consists of four steps in which the substrate is attached to the active serine, resulting in a tetrahedral intermediate that is stabilized by the catalytic His and Asp residues [125]. A tetrahedral intermediate is formed by the attack of a nucleophile, which following resolution produces the desired product (an acid or an ester) and free enzyme. The interfacial activation phenomenon allows lipases to be separated from esterases (which is only observed for lipases) [126]. A hydrophobic domain (lid) shielding the active site of lipase is responsible for this interfacial activation, according to structure elucidation. This lid will only open in the existence of the least concentration of substrate, such as a hydrophobic solvent that is organic in nature or a triglyceride phase, allowing access to the active site [127]. Esterases and lipases were among the main enzymes to be investigated for stability and activity in organic solvents; however, lipases exhibit this property more clearly. In the clothing industry, lipases are mostly utilized for bio-scouring, desizing fabrics, and surface functionalization of synthetic fibers. They may also be used in conjunction with other enzymes including protease, and xylanases [128]. Polyester fabrics' ability to absorb color is improved by the lipase enzyme, which also causes less surface abrasion and weight loss.

The fabrics made of PET (polyethylene terephthalate), PAN (polyacrylonitrile), and PA (polyamide) exhibit exceptional qualities such as good potency, great chemical endurance, minimal abrasion, and minimal shrinkage [129]. Extraordinary hydrophobicity and crystallinity, which impair wearing comfort (making these fabrics less suitable to be in contact with human skin), as well as the processing of fabrics (preventing the usage of finishing chemicals and dyeing representatives), are major drawbacks of synthetic fabrics [130]. Major finishing procedures/agents depend on water, hence increasing the hydrophilicity of the fabric surface is necessary. Currently, sodium hydroxide-based chemical treatments are utilized to increase the hydrophilicity and increase flexibility of fabrics [131]. Chemical treatment, which can cause undesirable weight and strength losses as well as irretrievable yellowing in the example of PA and PAN fabrics, is challenging to control [132]. The method also harms the environment because it uses a lot of energy and chemicals. Utilizing enzymes to adapt the surface of synthetic fabrics is a recently discovered substitute [133]. Substantial depilling, effective desizing, elevated hydrophilicity and reactivity with cationic dyes, and enhanced oily stain release were all outcomes of the enzymatic treatment [108].

5.8 Catalase

The enzymes that accelerate the breakdown of hydrogen peroxide into water and oxygen are referred to as catalases, also known as hydroperoxidases. After desizing and scouring but before dyeing, H_2O_2 is used in bleaching in the textile industry [134]. Historically, hydrogen peroxide was destroyed with a reducing agent and was rinsed with water. Conventionally, sodium bisulfate, which requires high temperatures and thorough rinsing, is used to eliminate hydrogen peroxide after bleaching [120]. Catalases can be used to break down extra hydrogen peroxide at low temperatures, this makes the procedure more affordable and environmentally friendly. They are made by a range of microorganisms, comprising bacteria and fungus, and many of them perform best at temperatures around 20°C and a pH of 07 [72]. The synthesis of microbial CATs will only be economically viable when using recombinant strains and low-cost technologies, or for CATs with particular qualities like thermostability or action at alkaline or acidic pH [135]. CATs from animal resources (bovine liver) are often inexpensive. The use of immobilized enzymes could lower the rate of enzyme for the breakdown of hydrogen peroxide in bleaching wastes, permitting not only the retrieval of the enzyme but also the reutilization of preserved decolorizing over-flows for coloring [136].

6. Future prospects

In practically every stage of manufacturing textile fibers, enzymes can be employed to create ecologically acceptable alternatives to chemical processes. Amylases for desizing, cellulases, laccases for denim finishing, and proteases included in commercial products are just a few examples of existing successful commercial uses. Commercial enzyme-based techniques for the bio-modification of artificial and natural fibers must first undergo additional study before they can be put into use. The quest for novel enzyme-producing microorganisms and enzymes derived from microorganisms is an important area of research. Future textile processing still has a lot of room for novel and enhanced enzyme uses.

7. Conclusion

Enzymes are the greatest substitute for the best textile processing they not only help the environment but also save a great deal of money by consuming less energy and water, which lowers the production cost. It appears that all processes will be able to be carried out utilizing enzymes in the future as the employment of diverse enzymes is still in its infancy, but their inventive uses are growing and expanding quickly into every aspect of textile production. Companies that produce enzymes are always working to make their products better for a wider variety of usage scenarios. The biggest barrier to employing enzymes is their high price. The textile industry was identified as a market area with great potential for implementing biotech but limited biotech awareness at the moment. Through the adoption and implementation of the enzymatic process, high value-added textile products with top quality may be produced while consuming less power, water, and other resources as well as enforcing to assure economic and environmental improvements along with sustainable development and social responsibility. In the area of treating textiles, enzymes are becoming much more prevalent and they can be used much more extensively in the textile industry if their expense can be controlled.

Author details

Wafa Haddar[1*], Shahid Adeel[2], Mahwish Salman[3*], Abdul Ghaffar[3], Mehwish Naseer[3], Muhammad Usama[3] and Manel Ben Ticha[1]

1 Department of Early Childhood, University College of Turbah, Taif University, Saudi Arabia

2 Department of Chemistry, Government College University, Faisalabad, Pakistan

3 Department of Biochemistry, Government College University, Faisalabad, Pakistan

*Address all correspondence to: wahaddar@tu.edu.sa and mahwishsalman@gcuf.edu.pk

References

[1] Chatha SAS, Asgher M, Iqbal H. Enzyme-based solutions for textile processing and dye contaminant biodegradation—a review. Environmental Science and Pollution Research. 2017;**24**(16):14005-14018. DOI: 10.1007/s11356-017-8998-1

[2] Gulzar T, Farooq T, Kiran S, Ahmad I, Hameed A. Green chemistry in the wet processing of textiles. In: The Impact and Prospects of Green Chemistry for Textile Technology. Woodhead Publishing; 2019. pp. 1-20. DOI: 10.1016/B978-0-08-102491-1.00001-0

[3] Kabir SMF, Chakraborty S, Hoque SMA, Mathur K. Sustainability assessment of cotton-based textile wet processing. Clean Technologies. 2019;**1**(1):232-246. DOI: 10.3390/cleantechnol1010016

[4] Madhav S, Ahamad A, Singh P, Mishra PK. A review of the textile industry: Wet processing, environmental impacts, and effluent treatment methods. Environmental Quality Management. 2018;**27**(3):31-41. DOI: 10.1002/tqem.21538

[5] Liyanapathiranage A, Peña MJ, Sharma S, Minko S. Nanocellulose-based sustainable dyeing of cotton textiles with minimized water pollution. ACS Omega. 2020;**5**(16):9196-9203. DOI: 10.1021/acsomega.9b04498

[6] Usmani Z, Sharma M, Sudheer S, Gupta VK, Bhat R. Engineered microbes for pigment production using waste biomass. Current Genomics. 2020;**21**(2):80-95. DOI: 10.2174/1389202921999200330152007

[7] Shen J. Enzymatic treatment of wool and silk fibers. Advances in Textile Biotechnology. 2nd ed. Elsevier; 2019. pp. 77-105. DOI: 10.1016/B978-0-08-102632-8.00005-0

[8] Shindhal T, Rakholiya P, Varjani S, Pandey A, Ngo HH, Guo W, et al. A critical review of advances in the practices and perspectives for the treatment of dye industry wastewater. Bioengineered. 2021;**12**(1):70-87. DOI: 10.1080/21655979.2020.1863034

[9] Sooch BS, Kauldhar BS, Puri M. Isolation and polyphasic characterization of a novel hyper catalase producing thermophilic bacterium for the degradation of hydrogen peroxide. Bioprocess and Biosystem Engineering. Germany. 2016;**39**(11):1759-1773. DOI: 10.1007/s00449-016-1651-4

[10] Singh D, Gupta N. Microbial laccase: A robust enzyme and its industrial applications. Biologia. 2020;**75**(8):1183-1193. DOI: 10.2478/s11756-019-00414-9

[11] Mojsov K, Andronikov D, Janevski A, Jordeva S, Kertakova M, Golomeova S, et al. Production and application of α-amylase enzyme in the textile industry. Tekstilna industrija. 2018;**66**(1):23-28

[12] Al-Ghanayem AA, Joseph B. Current prospective in using cold-active enzymes as eco-friendly detergent additive. Applied Microbiology and Biotechnology. 2020;**104**(7):2871-2882. DOI: 10.1007/s00253-020-10429-x

[13] Adeel S, Salman M, Zahoor AF, Usama M, Amin N. An insight into herbal-based natural dyes: isolation and applications. recycl from waste fash text a sustain circ econ approach. Willy; 2020.423-456. 10.1002/9781119620532.ch19

[14] Bouzidi A, Baaka N, Salem N, Mhenni MF, Mighri Z. Limoniastrum monopetalum stems as a new source of natural colorant for dyeing wool fabrics.

Fibers and Polymers. 2016;**17**(8):1256-1261. DOI: 10.1007/s12221-016-5664-z

[15] Atalah J, Cáceres-Moreno P, Espina G, Blamey JM. Thermophiles and the applications of their enzymes as new biocatalysts. Bioresource Technology. 2019;**280**:478-488. DOI: 10.1016/j.biortech.2019.02.008

[16] Shahid M, Mohammad F. Recent advancements in natural dye applications: a review. Journal of Cleaner Production. 2013;**53**:310-331. DOI: 10.1016/j.jclepro.2013.03.031

[17] Rajkumari NP, Dolakashoria S, Goswami P. Plant-based natural dye-stimulated visible-light reduction of go and physicochemical factors influencing the production of oxidizing species by a synthesized (rgo)/tio2 nanocomposite for environmental remediation. ACS omega. 2021;**6**(4):2686-2698. DOI: 10.1021/acsomega.0c04889

[18] Dixit A, Jain AK, Tiwari P, Gupta N, Gangele P. A phytopharmacological review on an important medicinal plant-Wrightia tinctoria. Current Research in Pharmaceutical Sciences. 2014;**4**(3):70-76. Available from: http://www.crpsonline.com/index.php/crps/article/view/125

[19] Hosseinnezhad M, Gharanjig K, Jafari R, Imani H, Razani N. Cleaner colorant extraction and environmentally wool dyeing using oak as eco-friendly mordant. Environmental Science and Pollution Research. 2021;**28**(6):7249-7260. DOI: 10.1007/s11356-020-11041-2

[20] Cooksey CJ. Quirks of dye nomenclature. 14. Madder: Queen of red dyes. Biotechnic and Histochemistry. 2020;**95**(6):474-482. DOI: 10.1080/10520295.2020.1714079

[21] Haddar W, Ben Ticha M, Meksi N, Guesmi A. Application of anthocyanins as natural dye extracted from Brassica oleracea L. var. capitata f. rubra: dyeing studies of wool and silk fibres. Natural Product Research. 2018;**32**(2):141-148. DOI: 10.1080/14786419.2017.1342080

[22] Gomashe SS, Ingle KP, Sarap YA, Chand D, Rajkumar S. Safflower (Carthamus tinctorius L.): An underutilized crop with potential medicinal values. Annals of Phytomedicine. 2021;**10**(1):242248. 10.21276/ap.2021.10.1.26

[23] Hosen MD, Rabbi MF, Raihan MA, Al Mamun MA. Effect of turmeric dye and biomordants on knitted cotton fabric coloration: A promising alternative to metallic mordanting. Cleaner Engineering Technology. 2021;**3**:1-11. DOI: 10.1016/j.clet.2021.100124

[24] Singh A, Sheikh J. Cleaner functional dyeing of wool using Kigelia Africana natural dye and Terminalia chebula bio-mordant. Sustainable Chemistry and Pharmacy. 2020;**17**:1-6. DOI: 10.1016/j.scp.2020.100286

[25] Singh A, Varghese LM, Battan B, Patra AK, Mandhan RP, Mahajan R. Eco-friendly scouring of ramie fibers using crude xylano-pectinolytic enzymes for textile purpose. Environmental Science and Pollution Research. Germany. 2020;**27**(6):6701-6710. DOI: 10.1007/s11356-019-07424-9

[26] Efstratiou E, Hussain AI, Nigam PS, Moore JE, Ayub MA, Rao JR. Antimicrobial activity of Calendula officinalis petal extracts against fungi, as well as Gram-negative and Gram-positive clinical pathogens. Complementary Therapy in Clinical Practice. 2012;**18**(3):173-176. DOI: 10.1016/j.ctcp.2012.02.003

[27] Dutta P, Mahjebin S, Sufian MA, Rabbi MR, Chowdhury S, Imran IH.

Impacts of natural and synthetic mordants on cotton knit fabric dyed with natural dye from onion skin in perspective of eco-friendly textile process. Materials Today Proceeding. 2021;**47**:2633-2640. DOI: 10.1016/j.matpr.2021.05.229

[28] Amutha K, Annapoorani SG, Sakthivel P, Sudhapriya N. Ecofriendly dyeing of textiles with natural dyes extracted from commercial food processing waste materials. Journal of Natural Fibers. 2022;**20**:1-8. DOI: 10.1080/15440478.2021.1993506

[29] Vankar PS, Shanker R, Srivastava J. Ultrasonic dyeing of cotton fabric with aqueous extract of Eclipta alba. Dyes and Pigments. 2007;**72**(1):33-37. DOI: 10.1016/j.dyepig.2005.07.013

[30] Gupta R, Gigras P, Mohapatra H, Goswami VK, Chauhan B. Microbial α-amylases: a biotechnological perspective. Process Biochemistry. 2003;**38**(11):1599-1616. DOI: 10.1016/S0032-9592(03)00053-0

[31] Sondhi S, Sharma P, Saini S, Puri N, Gupta N. Purification and characterization of an extracellular, thermo-alkali-stable, metal tolerant laccase from Bacillus tequilensis SN4. PLoS One. 2014;**9**(5):1-10. DOI: 10.1371/journal.pone.0096951

[32] Florczak T, Daroch M, Wilkinson MC, Białkowska A, Bates AD, Turkiewicz M, et al. Purification, characterisation and expression in Saccharomyces cerevisiae of LipG7 an enantioselective, cold-adapted lipase from the Antarctic filamentous fungus Geomyces sp. P7 with unusual thermostability characteristics. Enzyme and Microbial Technology. 2013;**53**(1):18-24. DOI: 10.1016/j.enzmictec.2013.03.021

[33] Ibrahim ASS, Al-Salamah AA, El-Badawi YB, El-Tayeb MA, Antranikian G. Detergent-, solvent-and salt-compatible thermoactive alkaline serine protease from halotolerant alkaliphilic Bacillus sp. NPST-AK15: purification and characterization. Extremophiles. 2015;**19**(5):961-971. DOI: 10.1007/s00792-015-0771-0

[34] Kurashiki R, Mizuno T, Murata K, Ohshiro T, Suzuki H. A plasmid vector that directs hyperproduction of recombinant proteins in the thermophiles Geobacillus species. Extremophiles. 2020;**24**(1):147-156. DOI: 10.1007/s00792-019-01142-3

[35] Cui L, Yuan J, Wang P, Wang X, Fan X, Wang Q. An ecofriendly phosphorylation of wool using Maillard reaction for improving cationic dye absorption. Journal of Cleaner Productiono. 2018;**178**:611-617. DOI: 10.1016/j.jclepro.2018.01.068

[36] Kim J-H, Kim D-H, So J-H, Koo H-J. Toward eco-friendly dye-sensitized solar cells (dsscs): natural dyes and aqueous electrolytes. Energies. 2021;**15**(1):1-18. DOI: 10.3390/en15010219

[37] Yusuf M, Shahid M, Khan SA, Khan MI, Islam S-U, Mohammad F, et al. Eco-dyeing of wool using aqueous extract of the roots of Indian madder (Rubia cordifolia) as natural dye. Journal of Natural Fibers. 2013;**10**(1):14-28. DOI: 10.1080/15440478.2012.738026

[38] Sk S, Mia R, Haque M, Shamim AM. Review on extraction and application of natural dyes. Text and Leather Reveiw. 2021;**4**(4):218-233. DOI: 10.3390/su14095599

[39] Swami C, Saini S, Gupta VB. Extraction of a natural dye from Sesbania aculeata plant. Journal of Textile Apparel, Technology and Managment. 2012;7(4):1-11

[40] Okonkwo SN, Ohanuzue CBC, Onuegbu GC, Obasi HC, Nnorom OO. Extraction of natural dyes from whitfieldia lateritia plant and its application on cotton fabric. Journal of Textile Science and Engineering. 2019;**9**(392):1-4. DOI: 10.4172/2165-8064.1000392

[41] Rather LJ, Ali A, Zhou Q, Ganie SA, Gong K, Rizwanul Haque QM, et al. Instrumental characterization of merino wool fibers dyed with Cinnamomum camphora waste/fallen leaves extract: An efficient waste management alternative. Journal of Cleaner Production. 2020;**273**:1-15. DOI: 10.1016/j.jclepro.2020.123021

[42] Kanniappan R. Assessment of dyeing properties and quality parameters of natural dye extracted from Lawsonia inermis. European Journal of Experimental Biology. London. 2015;**5**(7):62-70

[43] Rather LJ, Zhou Q, Ali A, Haque QMR, Li Q. Valorization of agro-industrial waste from peanuts for sustainable natural dye production: focus on adsorption mechanisms, ultraviolet protection, and antimicrobial properties of dyed wool fabric. ACS Food Science and Technology. 2021;**1**(3):427-442. DOI: 10.1021/acsfoodscitech.1c00005

[44] Aggarwal S. Indian dye yielding plants: Efforts and opportunities. In: Natural Resources Forum. Wiley Online Library; 2021. pp. 63-86. DOI: 10.1111/1477-8947.12214

[45] Kulkarni SS, Bodake UM, Pathade GR. Extraction of Natural Dye from Chili (Capsicum Annum) for Textile Coloration. Universal Journal of Environmental Resources and Technology. 2011;**1**(1):58-63

[46] Naveed T, Rehman F, Sanbhal N, Ali BA, Yueqi Z, Farooq O, et al. Novel natural dye extraction methods and mordants for textile applications. Surface Review and Letters. 2020;**27**(04):1-9. DOI: 10.1142/S0218625X1950135X

[47] Mirjalili M, Nazarpoor K, Karimi L. Eco-friendly dyeing of wool using natural dye from weld as co-partner with synthetic dye. Journal of Cleaner Production. 2011;**19**(9-10):1045-1051. DOI: 10.1016/j.jclepro.2011.02.001

[48] Zhang J, Sun L, Zhang H, Wang S, Zhang X, Geng A. A novel homodimer laccase from Cerrena unicolor BBP6: Purification, characterization, and potential in dye decolorization and denim bleaching. PLoS One. 2018;**13**(8):1-20. DOI: 10.1371/journal.pone.0202440

[49] Sivakumar V, Anna JL, Vijayeeswarri J, Swaminathan G. Ultrasound assisted enhancement in natural dye extraction from beetroot for industrial applications and natural dyeing of leather. Ultrasonic Sonochemistry. 2009;**16**(6):782-789. DOI: 10.1016/j.ultsonch.2009.03.009

[50] Sivakumar V, Vijaeeswarri J, Anna JL. Effective natural dye extraction from different plant materials using ultrasound. Industrial Crops and Products. 2011;**33**(1):116-122. DOI: 10.1016/j.indcrop.2010.09.007

[51] Sinha K, Das SP, Datta S. Response surface optimization and artificial neural network modeling of microwave assisted natural dye extraction from pomegranate rind. Industrial Crops and Products. 2012;**37**(1):408-414. DOI: 10.1016/j.indcrop.2011.12.032

[52] El-Apasery MA, Hussein AM, Nour El-Din NM, Saleh MO, El-Adasy A-BA. Microwave-assisted dyeing of wool fabrics with natural dyes as eco-friendly dyeing method: Part I. Dyeing

performance and fastness properties. Egyptian Journal of Chemistry. 2021;**64**(7):3751-3759. DOI: 10.21608/ejchem.2021.96598.4519

[53] Lo H, Chen C, Yang C, Wu P, Tsou C, Chiang K, et al. The carotenoid lutein enhances matrix metalloproteinase-9 production and phagocytosis through intracellular ROS generation and ERK1/2, p38 MAPK, and RARβ activation in murine macrophages. Journal of Leukocte Biology. 2013;**93**(5):723-735. DOI: 10.1189/jlb.0512238

[54] Comlekcioglu N, Efe L, Karaman S. Extraction of indigo from some Isatis species and dyeing standardization using low-technology methods. Brazilian Archives of Biology and Technolog. 2015;**58**:96-102. DOI: 10.1590/S1516-8913201502658

[55] Borges ME, Tejera RL, Díaz L, Esparza P, Ibáñez E. Natural dyes extraction from cochineal (Dactylopius coccus). New extraction methods. Food Chemistry. 2012;**132**(4):1855-1860. DOI: 10.1016/j.foodchem.2011.12.018

[56] Dhouibi N, Baaka N, Charradi R, Bouine I, Dhaouadi H, Dridi-Dhaouadi S. Multi-fiber dyeing improvement using natural supercritical CO_2 extracts. Fibers and Polymers. South Korea. 2021;**22**(7):1874-1882. DOI: 10.1007/s12221-021-0422-2

[57] Mukhopadhyay M. Extraction and processing with supercritical fluids. Journal of Chemical Technology and Biotechnolog. United Kingdom. 2009;**84**(1):6-12. DOI: 10.1002/jctb.2072

[58] Banchero M. Supercritical fluid dyeing of synthetic and natural textiles–a review. Color Technology. 2013;**129**(1):2-17. DOI: 10.1111/cote.12005

[59] Rahman M, Hack-Polay D, Billah MM, Un Nabi MN. Bio-based textile processing through the application of enzymes for environmental sustainability. International Journal of Technology Management and Sustainable Development. 2020;**19**(1):87-106. DOI: 10.1386/tmsd_00017_1

[60] Boyce A, Walsh G. Expression and characterisation of a thermophilic endo-1, 4-β-glucanase from Sulfolobus shibatae of potential industrial application. Molecular Biology Reports. 2018;**45**(6):2201-2211. DOI: 10.1007/s11033-018-4381-7

[61] Bhattacharya A, Goyal N, Gupta A. Degradation of azo dye methyl red by alkaliphilic, halotolerant Nesterenkonia lacusekhoensis EMLA3: application in alkaline and salt-rich dyeing effluent treatment. Extremophiles. 2017;**21**(3):479-490. DOI: 10.1007/s00792-017-0918-2

[62] Brininger C, Spradlin S, Cobani L, Evilia C. The more adaptive to change, the more likely you are to survive: Protein adaptation in extremophiles. Seminars in Cell & Developmental Biology. 2018;**84**:158-169. DOI: 10.1016/j.semcdb.2017.12.016

[63] Cabrera MÁ, Blamey JM. Biotechnological applications of archaeal enzymes from extreme environments. Biological Research. 2018;**51**:1-15. DOI: 10.1186/s40659-018-018-3

[64] Schultz J, Rosado AS. Extreme environments: a source of biosurfactants for biotechnological applications. Extremophiles. 2020;**24**(2):189-206. DOI: 10.1007/s00792-019-01151-2

[65] Shahid M, Mohammad F, Chen G, Tang R-C, Xing T. Enzymatic processing of natural fibres: white biotechnology for sustainable development. Green Chemistry. 2016;**18**(8):2256-2281. DOI: 10.1039/C6GC00201C

[66] Chen X-H, Li F, Li F-N, Chen M-S, Yan X-R, He Z-B, et al. Nocardioides acrostichi sp. nov., a novel endophytic actinobacterium isolated from leaf of Acrostichum aureum. Antonie Van Leeuwenhoek. S. 2021;**114**(4):479-486. DOI: 10.1007/s10482-021-01535-5

[67] Gil-Durán C, Ravanal M-C, Ubilla P, Vaca I, Chávez R. Heterologous expression, purification and characterization of a highly thermolabile endoxylanase from the Antarctic fungus Cladosporium sp. Fungal Biology. 2018;**122**(9):875-882. DOI: 10.1016/j.funbio.2018.05.002

[68] Joshi N, Sharma M, Singh SP. Characterization of a novel xylanase from an extreme temperature hot spring metagenome for xylooligosaccharide production. Applied Microbiology and Biotechnology. 2020;**104**(11):4889-4901. DOI: 10.1007/s00253-020-10562-7

[69] Kalia S, Sheoran R. Modification of ramie fibers using microwave-assisted grafting and cellulase enzyme–assisted biopolishing: a comparative study of morphology, thermal stability, and crystallinity. International Journal of Polymer Analysis and Characterization. 2011;**16**(5):307-318. DOI: 10.1080/1023666X.2011.587946

[70] Islam M, Nahar K, Ferdush J, Akter T. Impact of bleaching actions of bleaching powder and hydrogen peroxide on biopolished denim garments. Tekstil. 2019;**68**(1-3):35-39

[71] Chand N, Nateri AS, Sajedi RH, Mahdavi A, Rassa M. Enzymatic desizing of cotton fabric using a Ca2+-independent α-amylase with acidic pH profile. Journal of Molecular Catalysis B Enzymatic. 2012;**83**:46-50. DOI: 10.1016/j.molcatb.2012.07.003

[72] Mechri S, Bouacem K, Zaraî Jaouadi N, Rekik H, Ben Elhoul M, Omrane Benmrad M, et al. Identification of a novel protease from the thermophilic Anoxybacillus kamchatkensis M1V and its application as laundry detergent additive. Extremophiles. 2019;**23**(6):687-706. DOI: 10.1007/s00792-019-01123-6

[73] Kumar D, Kumar A, Sondhi S, Sharma P, Gupta N. An alkaline bacterial laccase for polymerization of natural precursors for hair dye synthesis. 3 Biotech. 2018;**8**(3):1-10. DOI: 10.1007/s13205-018-1181-7

[74] Kumar JA, Kumar MS. A study on improving dyeability of polyester fabric using lipase enzyme. Autex Research Journal. 2020;**20**(3):243-249. DOI: 10.2478/aut-2019-0030

[75] Klanovicz N, Camargo AF, Stefanski FS, Zanivan J, Scapini T, Pollon R, et al. Advanced oxidation processes applied for color removal of textile effluent using a home-made peroxidase from rice bran. Bioprocess and Biosystem Engineering. 2020;**43**(2):261-272. DOI: 10.1007/s00449-019-02222-6

[76] Aggarwal R, Dutta T, Sheikh J. Extraction of amylase from the microorganism isolated from textile mill effluent vis a vis desizing of cotton. Sustainable Chemistry and Pharmacy. 2019;**14**:1-6. DOI: 10.1016/j.scp.2019.100178

[77] Aggarwal R, Dutta T, Sheikh J. Extraction of pectinase from Candida isolated from textile mill effluent and its application in bio-scouring of cotton. Sustainable Chemistry and Pharmacy. 2020;**17**:1-7. DOI: 10.1016/j.scp.2020.100291

[78] Saha P, Khan MF, Patra S. Truncated α-amylase: an improved candidate for textile processing. Preprative Biochemistry and Biotechnology.

2018;**48**(7):635-645. DOI: 10.1080/10826068.2018.1479863

[79] Nair HP, Bhat SG. Arabian Sea metagenome derived-α-amylase P109 and its potential applications. Ecological Genetics and Genomics. 2020;**16**:1-5. DOI: 10.1016/j.egg.2020.100060

[80] Asoodeh A, Alemi A, Heydari A, Akbari J. Purification and biochemical characterization of an acidophilic amylase from a newly isolated Bacillus sp. DR90. Extremophiles. 2013;**17**(2):339-348. DOI: 10.1007/s00792-013-0520-1

[81] Adeel S, Rehman FU, Zia KM, Azeem M, Kiran S, Zuber M, et al. Microwave-supported green dyeing of mordanted wool fabric with arjun bark extracts. Journal of Natural Fibers. 2021;**18**(1):136-150. DOI: 10.1080/15440478.2019.1612810

[82] Ferrari M, Mazzoli R, Morales S, Fedi M, Liccioli L, Piccirillo A, et al. Enzymatic laundry for old clothes: immobilized alpha-amylase from Bacillus sp. for the biocleaning of an ancient Coptic tunic. Applied Microbiology and Biotechnology. 2017;**101**(18):7041-7052. DOI: 10.1007/s00253-017-8437-8

[83] Arabacı N, Arıkan B. Isolation and characterization of a cold-active, alkaline, detergent stable α-amylase from a novel bacterium *Bacillus subtilis* N8. Preparative Biochemistry and Biotechnology. 2018;**48**(5):419-426. DOI: 10.1080/10826068.2018.1452256

[84] Ali I, Akbar A, Anwar M, Prasongsuk S, Lotrakul P, Punnapayak H. Purification and characterization of a polyextremophilic α-amylase from an obligate halophilic *Aspergillus penicillioides* isolate and its potential for souse with detergents. Biomed Research International. 2015;**2015**:245649. DOI: 10.1155/2015/245649

[85] Mojsov K. Enzymatic desizing, bioscouring and enzymatic bleaching of cotton fabric with glucose oxidase. Journal of Textile Instustry. 2019;**110**(7):1032-1041. DOI: 10.1080/00405000.2018.1535240

[86] Sankarraj N, Nallathambi G. Enzymatic biopolishing of cotton fabric with free/immobilized cellulase. Carbohydrate Polymers. 2018;**191**:95-102. DOI: 10.1016/j.carbpol.2018.02.067

[87] Pušić T, Tarbuk A, Dekanić T. Bio-innovation in cotton fabric scouring-acid and neutral pectinases. Fibres and Textile in Eastern Europe. 2015;**109**(1):98-103. Available from: https://urn.nsk.hr/urn:nbn:hr:201:364654

[88] Nov Z, Biobeljenja E. New combined bio-scouring and bio-bleaching process of cotton fabrics. Materials Technology. 2013;**47**:409-412

[89] Kumari A, Kishor N, Guptasarma P. Characterization of a mildly alkalophilic and thermostable recombinant Thermus thermophilus laccase with applications in decolourization of dyes. Biotechnology Letters. 2018;**40**(2):285-295. DOI: 10.1007/s10529-017-2461-8

[90] Chen K, Mo Q, Liu H, Yuan F, Chai H, Lu F, et al. Identification and characterization of a novel cold-tolerant extracellular protease from Planococcus sp. CGMCC 8088. Extremophiles. 2018;**22**(3):473-484. DOI: 10.1007/s00792-018-1010-2

[91] Carrasco M, Rozas JM, Alcaíno J, Cifuentes V, Baeza M. Pectinase secreted by psychrotolerant fungi: identification, molecular characterization and heterologous expression of a cold-active polygalacturonase from Tetracladium sp. Microbial Cell Factories. 2019;**18**(1):1-11. DOI: 10.1186/s12934-019-1092-2

[92] Kalantzi S, Kekos D, Mamma D. Bioscouring of cotton fabrics by

multienzyme combinations: application of Box–Behnken design and desirability function. Cellulose. 2019;**26**(4):2771-2790. DOI: 10.1007/s10570-019-02272-9

[93] Mate DM, Alcalde M. Laccase engineering: from rational design to directed evolution. Biotechnology Advances. 2015;**33**(1):25-40. DOI: 10.1016/j.biotechadv.2014.12.007

[94] Panwar V, Sheikh JN, Dutta T. Sustainable denim bleaching by a novel thermostable bacterial laccase. Applied Biochemistry and Biotechnology. 2020;**192**(4):1238-1254. DOI: 10.1007/s12010-020-03390-y

[95] Rodríguez-Couto S. A promising inert support for laccase production and decolouration of textile wastewater by the white-rot fungus Trametes pubescesns. Journal of Hazardous Materials. 2012;**233**:158-162. DOI: 10.1016/j.jhazmat.2012.07.003

[96] Sondhi S, Kumar D, Angural S, Sharma P, Gupta N. Enzymatic Approach for Bioremediation of Effluent from Pulp and Paper Industry by Thermo Alkali Stable Laccase from Bacillus Tequilensis SN4. Journal of Commercial Biotechnology. 2017;**23**(4):12-22 link.gale.com/apps/doc/A664819298/AONE?

[97] Sondhi S, Kaur R, Kaur S, Kaur PS. Immobilization of laccase-ABTS system for the development of a continuous flow packed bed bioreactor for decolorization of textile effluent. Internatioonal Journal of Biological Macromolecules. 2018;**117**:1093-1100. DOI: 10.1016/j.ijbiomac.2018.06.007

[98] Unuofin JO. Treasure from dross: application of agroindustrial wastes-derived thermo-halotolerant laccases in the simultaneous bioscouring of denim fabric and decolorization of dye bath effluents. Industrials Crops and Products. 2020;**147**:1-8. DOI: 10.1016/j.indcrop.2020.112251

[99] Tülek A, Karataş E, Çakar MM, Aydın D, Yılmazcan Ö, Binay B. Optimisation of the production and bleaching process for a new Laccase from Madurella mycetomatis, expressed in Pichia pastoris: From secretion to yielding prominent. Molecular Biotechnology. 2021;**63**(1):24-39. DOI: 10.1007/s12033-020-00281-9

[100] Solanki P, Putatunda C, Kumar A, Bhatia R, Walia A. Microbial proteases: ubiquitous enzymes with innumerable uses. 3. Biotech. 2021;**11**(10):1-25. DOI: 10.1007/s13205-021-02928-z

[101] Tapia-Tussell R, Pereira-Patrón A, Alzate-Gaviria L, Lizama-UcG, Pérez-BritoD, Solis-PereiraS. Decolorization of textile effluent by Trametes hirsuta Bm-2 and lac-T as possible main laccase-contributing gene. Current Microbiolog. 2020;**77**(12):3953-3961. DOI: 10.1007/s00284-020-02188-9

[102] Basheer S, Rashid N, Ashraf R, Akram MS, Siddiqui MA, Imanaka T, et al. Identification of a novel copper-activated and halide-tolerant laccase in Geobacillus thermopakistaniensis. Extremophiles. 2017;**21**(3):563-571. DOI: 10.1007/s00792-017-0925-3

[103] Andreaus J, Olekszyszen DN, Silveria MHL. Processing of cellulosic textile materials with cellulases. Cellulose and other Natural Occuring Polymers. 2014; Research Signpost 37/661 (2), Kerala, India. pp. 11-19. B

[104] Wang R, Yang C, Fang K, Cai Y, Hao L. Removing the residual cellulase by graphene oxide to recycle the bio-polishing effluent for dyeing cotton fabrics. Journal of Environmental Management. 2018;**207**:423-431. DOI: 10.1016/j.jenvman.2017.11.056

[105] Hu F, Jung S, Ragauskas A. Pseudo-lignin formation and its impact

on enzymatic hydrolysis. Bioresource Technology. 2012;**117**:7-12. DOI: 10.1016/j.biortech.2012.04.037

[106] Šimić K, Soljačić I, Pušić T. Application of cellulases in the process of finishing. Tekstilec. 2015;**58**(1):47-56. DOI: 10.14502/Tekstilec2015.58.47%E2%88%9256

[107] Mojsov K. Microbial cellulases and their applications in textile processing. International Journal of Marketing and Technology. 2012;**2**(11):12-29. Available from: https://eprints.ugd.edu.mk/id/eprint/1596

[108] Yu Y, Yuan J, Wang Q, Fan X, Wang P, Cui L. A study of surface morphology and structure of cotton fibres with soluble immobilized-cellulase treatment. Fibers and Polymers. 2014;**15**(8):1609-1615. DOI: 10.1007/s12221-014-1609-6

[109] Gaonkar SK, Furtado IJ. Characterization of extracellular protease from the haloarcheon Halococcus sp. strain GUGFAWS-3 (MF425611). Current Microbiology. 2020;**77**(6):1024-1034. DOI: 10.1007/s00284-020-01896-6

[110] Manikandan M, Pašić L, Kannan V. Purification and biological characterization of a halophilic thermostable protease from Haloferax lucentensis VKMM 007. World Journal of Microbiology and Biotechnology. 2009;**25**(12):2247-2256. DOI: 10.1007/s11274-009-0132-1

[111] Begum S, Wu J, Takawira CM, Wang J. Surface modification of polyamide 6, 6 fabrics with an alkaline protease–subtilisin. Journal of Engineering Fibers and Fabrics. 2016;**11**(1):1-11. 10.1177/155892501601100110

[112] Waly AI, Marie MM, Shahin MF, Faroun NMS. Effect of protease treatment on the physical properties and dyeability of wool/nylon blend to cutch natural dye. International Journal of Science and Research. 2016;**5**(2):1764-1770

[113] Ibrahim NA, El-Shafei HA, Abdel-Aziz MS, Ghaly MF, Eid BM, Hamed AA. The potential use of alkaline protease from Streptomyces albidoflavus as an eco-friendly wool modifier. Journal of Textile Instustries. 2012;**103**(5):490-498. DOI: 10.1080/00405000.2011.588417

[114] Oyeleke SB, Egwim EC, Auta SH. Screening of Aspergillus flavus and Aspergillus fumigatus strains for extracellular protease enzyme production. Journal of Microbiology and Antimicrobials. 2010;**2**(7):83-87

[115] Gimenes NC, Silveira E, Tambourgi EB. An overview of proteases: production, downstream processes and industrial applications. Seperation and Purification Reviews. 2021;**50**(3):223-243. DOI: 10.1080/15422119.2019.1677249

[116] Chen S, Gao H, Chen J, Wu J. Surface modification of polyacrylonitrile fibre by nitrile hydratase from Corynebacterium nitrilophilus. Applied Biochemistry and Biotechnology. 2014;**174**(6):2058-2066. DOI: 10.1007/s12010-014-1186-6

[117] Shen J-D, Cai X, Liu Z-Q, Zheng Y-G. Nitrilase: a promising biocatalyst in industrial applications for green chemistry. Critical Reviews in Biotechnology. 2021;**41**(1):72-93. DOI: 10.1080/07388551.2020.1827367

[118] Nigam VK, Arfi T, Kumar V, Shukla P. Bioengineering of nitrilases towards its use as green catalyst: applications and perspectives. Indian Journal of Microbiology. 2017;**57**(2):131-138. DOI: 10.1007/s12088-017-0645-5

[119] Akkaya A, Ozseker EE. Modification of polyacrylonitrile fabric for antibacterial application by tetracycline immobilization. Polymer Testing. 2019;**78**:1-11. DOI: 10.1016/j.polymertesting.2019.105959

[120] Easson M, Condon B, Villalpando A, Chang S. The application of ultrasound and enzymes in textile processing of greige cotton. Ultrasonics. 2018;**84**:223-233. DOI: 10.1016/j.ultras.2017.11.007

[121] Kaplan O, Nikolaou K, Pišvejcová A, Martínková L. Hydrolysis of nitriles and amides by filamentous fungi. Enzyme and Microbial Technology. 2006;**38**(1-2):260-264. DOI: 10.1016/j.enzmictec.2005.07.022

[122] Vejvoda V, Kaplan O, Bezouška K, Martínková L. Mild hydrolysis of nitriles by the immobilized nitrilase from Aspergillus niger K10. Journal of Molecular Catalysis B Enzymatics. 2006;**39**(1-4):55-58. DOI: 10.1016/j.molcatb.2006.01.027

[123] Okamoto S, Eltis LD. Purification and characterization of a novel nitrile hydratase from Rhodococcus sp. RHA1. Molecular Microbiology. 2007;**65**(3):828-838. DOI: 10.1111/j.1365-2958.2007.05834.x

[124] El-Shemy NS, El-Hawary NS, El-Sayed H. Basic and reactive-dyeable polyester fabrics using lipase enzymes. Journal of Chemical Engineering Process Technology. 2016;**7**(1):1-5. DOI: 10.4172/2157-7048.1000271

[125] Mukendi GM, Mitema A, Nelson K, Feto NA. Bacillus Species of Ruminant Origin as a Major Potential Sources of Diverse Lipolytic Enzymes for Industrial and Therapeutic Applications. In: Bacilli in Agrobiotechnology. Springer; 2022. pp. 255-283

[126] Patel R, Trivedi U, Patel KC. Lipases: An efficient biocatalyst for biotechnological applications. Journal of Microbiology and Biotechnology Food and Sciences. Slovakia. 2022;**11**(5):1-7. DOI: 10.55251/jmbfs.2498

[127] Angelin J, Kavitha M. Extremophilic Fungal Lipases: Screening, Purification, Assay, and Applications. In: Extremophilic Fungi. Springer; 2022. pp. 395-438. DOI: 10.1007/978-981-16-4907-3_18

[128] Khan MF, Kundu D, Hazra C, Patra S. A strategic approach of enzyme engineering by attribute ranking and enzyme immobilization on zinc oxide nanoparticles to attain thermostability in mesophilic Bacillus subtilis lipase for detergent formulation. International Journal of Biological Macromolecules. 2019;**136**:66-82. DOI: 10.1016/j.ijbiomac.2019.06.042

[129] Ortiz C, Ferreira ML, Barbosa O, dos Santos JCS, Rodrigues RC, Berenguer-Murcia Á, et al. Novozym 435: the "perfect" lipase immobilized biocatalyst. Catalysis Science and Technology. United Kingdom. 2019;**9**(10):2380-2420. DOI: 10.1039/C9CY00415G

[130] Ismail AR, Kashtoh H, Baek K-H. Temperature-resistant and solvent-tolerant lipases as industrial biocatalysts: Biotechnological approaches and applications. International Journal of Biological Macromolecules. 2021;**187**:127-142. DOI: 10.1016/j.ijbiomac.2021.07.101

[131] Vivek K, Sandhia GS, Subramaniyan S. Extremophilic lipases for industrial applications: A general review. Biotechnology Advances. 2022;**60**:1-11. DOI: 10.1016/j.biotechadv.2022.108002

[132] de Alencar Figueira Angelotti J, Battaglia Hirata D, Rodrigues dos Santos M. Lipases: Sources of

acquisition, ways of production, and recent applications. Catalysis Research. 2022;**2**(2):1-9. DOI: 10.21926/cr.2202013

[133] Barbosa O, Ruiz M, Ortiz C, Fernández M, Torres R, Fernandez-Lafuente R. Modulation of the properties of immobilized CALB by chemical modification with 2, 3, 4-trinitrobenzenesulfonate or ethylendiamine. Advantages of using adsorbed lipases on hydrophobic supports. Process Biochemistry. 2012;**47**(5):867-876. DOI: 10.1016/j.procbio.2012.02.026

[134] Santos A, Mendes S, Brissos V, Martins LO. New dye-decolorizing peroxidases from Bacillus subtilis and Pseudomonas putida MET94: towards biotechnological applications. Applied Microbiology and Biotechnology. 2014;**98**(5):2053-2065. DOI: 10.1007/s00253-013-5041-4

[135] Zafar A, Aftab MN, Iqbal I, ud Din Z, Saleem MA. Pilot-scale production of a highly thermostable α-amylase enzyme from Thermotoga petrophila cloned into E. coli and its application as a desizer in the textile industry. RSC. Advances. 2019;**9**(2):984-992. DOI: 10.1039/C8RA06554C

[136] Rekik H, Jaouadi NZ, Bouacem K, Zenati B, Kourdali S, Badis A, et al. Physical and enzymatic properties of a new manganese peroxidase from the white-rot fungus Trametes pubescens strain i8 for lignin biodegradation and textile-dyes biodecolorization. International Journal of Biological Macromolecules. 2019;**125**:514-525. DOI: 10.1016/j.ijbiomac.2018.12.053

Chapter 4

Comparison and Analysis of Colorant in Toner Cartridges: A Material Safety Data Sheet Study

Kelvin Sin-Wang Kwan and Chi-Wai Kan

Abstract

Nowadays, electronic devices such as mobile phones and tablets are gadgets that have become common in most people's daily lives. Although people often use tablets to read files, newspapers, or other papers, printing some documents is still necessary and convenient for most people. Therefore, a printer is one of the basic machines that many people use for work or to learn. Toner cartridges are the main components that print high-quality images or text on paper and are therefore of research value. Existing literature lacks research on performance of different printers and toners. Therefore, this study investigated and analyzed different types and brands of toners. In the study, toner cartridges provided by the four major suppliers were compared with data provided in material safety data sheets (MSDS) based primarily on different products. A comprehensive review and analysis of the concentration, function, definition, and impact involved in the product was conducted in this study.

Keywords: toner, dyes, pigment, MSDS, health

1. Introduction

The toner cartridge is a part of a laser printer and it contains a complex powder used for printing images and text on paper. Besides laser printers, it is also used in compound printers and fax machines based on electronic photography technology. The chemicals in a toner are a mixture of colored pigments, plastic resins, and other ingredients.

Empty toner cartridges are a waste and their reuse and recycling have attracted much attention in recent years. Printers and toner cartridges have serious problems, from the manufacturing process to the eventual throw away and recycling of materials. Most chemicals in toner cartridges have the potential to cause damage to the environment [1–3] and humans [4–7]. Discarded toner cartridges may cause damage to the environment as they can contaminate soil and water when sent to landfills. In fact, carbon black in toner is classified as a potential carcinogen because the devastating effects of toner cannot be properly addressed [8].

In order to compare and analyze different contents of toner cartridges, this study selected four well-known brands that are sharing key markets, based on the global printer market share of suppliers from 2015 to 2018. The main objectives of the study are:

1. To analyze components of toners and compare the toner cartridges of the four brands based on the material safety data sheet (MSDS).

2. To study the hazardous impacts to the environment and human health caused by waste toner cartridges.

2. Methodology

2.1 Selection of toner cartridge brands

Based on Staista 2019 [9], four printer brands (Labeled Brand A, Brand B, Brand C, and Brand D) together having a total market share in excess of 50% between 2015 and 2018 were selected for the study. Basically, printer sales are directly related to toner cartridge sales because they constitute the relationship between major products and sub-products.

2.2 Material safety data sheet

The material safety data sheet (MSDS) is a document that provides safety and health measures that need to be followed when using different components of the product(s). It is a commonly used system for listing information about chemicals, compounds, and chemical mixtures. The MSDS contains instructions for safe use and potential hazards associated with a particular product or material. The composition of toner and its potentially dangerous effects on human health can be found in MSDS. Typically, each toner cartridge has a MSDS that is included in the package or displayed on its official website. Most of the data in this study was obtained from MSDS.

2.2.1 Procedure of collecting data from material safety data sheet

1. All MSDS of toner cartridges of each brand were obtained from official websites of the brands.

2. MSDS was studied and the percentage of each component in each toner cartridge was recorded.

2.3 Search the health and environment impacts based on collected data

Names of different chemicals contained in toners are listed in MSDS and the precautions that need to be taken for avoiding adverse health effects are briefly described. Based on the data collected, these preventive measures are analyzed to determine health and environmental impacts.

3. Result and discussion

3.1 Result of data collected from MSDS

Data were collected for 1473 toner cartridges. As shown in **Table 1**, there were 668 cartridges of Brand A, 353 of Brand B, 296 of Brand C, and 156 cartridges of Brand D. In **Table 1**, the data show that black color accounts for about 50% of the total for each brand, while the other three colors account for about 15% each, of the total. In daily use, black color is the main color used in work, study, or printing. As a result, most companies have introduced black color as the main type of toner.

3.2 Analysis of the main chemicals in toner cartridges

Ingredients in all toner brands are not the same, though they are similar. **Table 2** lists all chemicals in toner cartridges of the four chosen brands. Brand A cartridges contain 42 types of chemicals. In addition, Brand A contains most of the chemicals in all the four brands. Only Brand A contains green, red, and translucent white colors in addition to the four main colors. Brand B has 17 types, Brand C has 21 types and Brand D has 20 types. Cyan, magenta, yellow, and black are the four main colors in different toners.

The numbers in brackets indicate the total number of cartridges in which the particular chemical was found.

Brand A (668 cartridges) contained 42 types of chemicals; the highest number of chemicals used by one brand in the four Brands. The main chemicals in Brand A cartridges were glycerin and water inside the toner, found in 475 and 496 cartridges, respectively, or over 70% of cartridges of the same model. Other chemicals found were 1,2-benzisothiazol-3(2H)-one, 2-Pyrrolidinone and glycol, found in 261, 224, and 292 cartridges, respectively.

Brand B is using 17 types of chemicals; 353 cartridges were examined. This was the least number of types compared with the other three brands. Basically, there are six types of chemicals used in over 50% of the cartridges: carbon black, fatty acid ester, PMMA, silicon dioxide, styrene-acrylate copolymer, and styrene-acrylate resin.

Brand C is using 21 types of chemicals in 296 toner cartridges, in which amorphous silica, styrene acrylate copolymer, and wax are the most used chemicals, found in 215, 255, and 220 cartridges, respectively.

Lastly, Brand D uses four main types of chemicals: carbon black, paraffin wax, polyester resin, and silica. Obviously, the ingredients are slightly different from the

Brand	No. of Model	*Type of Chemical	Cyan Color	Magenta Color	Yellow Color	Black Color	**Other Colors
A	668	42	116	117	90	311	34
B	353	17	36	36	36	245	0
C	296	21	54	54	55	133	0
D	156	20	18	20	19	99	0

**Type of chemical is the total amount of chemical used in the brand. **Some cartridges are classified as having other colors such as green and red.*

Table 1.
Basic information.

Brand A
1,2-benzisothiazol-3(2H)-one (261); 2-Pyrrolidinone (224); 3-Methyl-1,5-pentanediol (17); Ammonia derivative (40); Ammonium benzoate (7); Amorphous silica (130); Black pigment (1); Calcium nitrate (1); Carbon black (72); Diethylene glycol (73); Ethylene glycol (Glycol) (79); Ethylene urea (137); FD&C Yellow No. 6 (2); Ferrite (14); Ferrite including manganese (8); Ferrite including Zinc (12); Glycerin (475); Glycol (292); Iron oxide (30); Isopropyl alcohol (64); Lactam (73); Magenta dye (9); Magenta pigment (8); Magnesium nitrate (9); Pigment (80); Polyester resin (31); Pyridine azo dye (22); Pyridone dye (2); Quinoline derivative (10); Styrene polymer (3); Styrene-acrylate copolymer (139); Substituted anthraquinone derivative (8); Substituted naphthalene sulfonic acid (11); Substituted naphthol azo dye (4); Substituted phthalocyanine salt (43); Thiadiazole azo dye (7); Thiodiglycol (8); Titanium dioxide (34); Triazine derivative (6); Triol (15); Water (496) Wax (104)
Brand B
1,2-benzisothiazol-3(2H)-one (1); Carbon Black (bound) (245); Diethylene glycol (1); Fatty Acid Ester (225); Glycerol (1); Paraffin wax (40); Pigment (108); Pigment – 1 (19); Pigment – 2 (19); PMMA (255); Polyester resin (98); Rosin, fumarate (51); Silicon dioxide (amorphous) - 112,945-52-5 (218); Silicon dioxide (amorphous) - 844,491-94–7 (165); Styrene-acrylate copolymer (254); Styrene-acrylate resin (181); Water (1)
Brand C
Aluminum alloy (1); Amorphous silica (215); Aryl amine derivative (1); Carbon black (61); Ceramic materials (9); Copper compound (14); Coating materials (4); Cyan pigment (4); Ferrite (21); Ferrite including zinc (2); Iron oxide (56); Magenta pigment (4); Paraffin wax (22); Pigment (141); Polyester resin (54); Silica (16); Silicon dioxide (5); Styrene acrylate copolymer (255); Titanium dioxide (80); Wax (220); Yellow pigment (4)
Brand D
Carbon black (108); Carnauba wax (27); Ceramic material (29); Coating materials (24); Cyan pigment (18); Ester wax (2); Ethylene wax (33); Ferrite (3); Magenta pigment (20); Paraffin wax (122); Pigment (9); Polyester (3); Polyester resin (100); Silica (135); Silicon dioxide (5); Styrene-acrylic resin (53); Tin (IV) oxide (2); Titanium dioxide (38); Wax (6); Yellow pigment (19)

Table 2.
All chemicals list.

other three brands. The main ingredient is a chemical compound, which is, in all the four brands, the main ingredient.

Basically, all brands of toner cartridges contain similar but different ingredients. Different chemicals can have the same function in the printing process. For example, iron oxide and silica are completely different in physical and chemical properties, but they have similar functions in a toner. **Table 3** shows chemicals present in these target brands. In particular, carbon black, polyester resin, and silica are the only three chemicals that are present in each brand.

The value of carbon black is between 10.8 and 69.4%. The quantity and percentage indicate that carbon black is the main chemical in these toners. Brand B and Brand D are relatively high in carbon black because they contain about 70% carbon black and are valued on the color capacity of toner. Because carbon black is the main ingredient in the black toner cartridge, most toner brands use it as the primary source of color.

Polyester resin is used in different ratios in these four brands because it is used as a toner binder in laser printers. The percentage of polyester resins is between 4.6 and 64.1%. Brand A contains 4.6% polyester resin, which is not the main component of the control binding and release properties. By contrast, Brand D has 64.1% polyester resin because it relies on one chemical to produce high mechanical strength and excellent viscous elasticity.

DOI: http://dx.doi.org/10.5772/intechopen.107439

Brand	A		B		C		D	
Chemical	**Amount**	**Percentage***	**Amount**	**Percentage***	**Amount**	**Percentage***	**Amount**	**Percentage***
Carbon black	72	10.8%	245	69.4%	61	20.6%	108	69.2%
Polyester resin	31	4.6%	98	27.8%	54	18.2%	100	64.1%
Titanium dioxide	34	5.1%	0	0%	80	27.0%	38	24.4%
Paraffin wax	0	0%	40	11.3%	22	7.4%	122	78.2%
Wax	104	15.6%	0	0%	220	74.3%	6	3.8%
Silica	130	19.5%	0	0%	16	5.4%	135	86.5%
Silicon dioxide	0	0%	232**	65.7%	5	1.7%	5	3.2%
Iron oxide	30	4.5%	0	0%	56	18.9%	0	0%
Ferrite	34***	5.1%	0	0%	23	7.8%	3	1.9%

**The amount of material divided by total number of cartridges. **Includes all silicon dioxide in different cartridges. ***Includes all ferrite in different metals.*

Table 3.
Same materials between four brands.

Silicon dioxide, silica, iron oxide, and ferrite have essentially the same function in the process of printing. In this study, the percentage of these chemicals was found to be between 29.1 and 91.6%. Brand D's percentage is considered satisfactory because most models use these chemicals as carriers to bind the right position of the rotating drum.

3.3 Discussion of chemicals with high percentages in MSDS

Most of the chemicals collected are represented as a percentage range. Percentage refers to the concentration of chemicals in toner, which impacts efficacy in terms of high or low values. This section discusses major chemicals with a high percentage as stated in the MSDS.

3.3.1 Carbon black

As shown in **Table 4**, the percentage of carbon black in toner cartridges of the four brands is in the range of 1–10%. Obviously, carbon black is not the main ingredient of toner as it only provides color. However, it is still the main content used for coloring in every brand. In Brand A cartridges, carbon black content is only in three different ranges, which is the smallest number, and the values of the three types of percentages are more uniform. In contrast, Brand D has different percentage ranges, with 16 types found in MSDS, with primary concentrations ranging from <1 to <5%.

[1] Brand A		Brand B		Brand C		Brand D	
Concentration	**Amount**	**Concentration**	**Amount**	**Concentration**	**Amount**	**Concentration**	**Amount**
< 1%	24	1–5%	16	< 1%	4	0.1%	3
1–5%	22	1–10%	1	< 2%	4	0.3.%	1
5–10%	26	2.5–5.5%	61	< 6%	14	1–5%	1
		3–10%	21	< 8%	4	1–6%	20
		4–7%	25	< 10%	33	1–10%	1
		4–14%	11	< 15%	2	2–5%	2
		5–7%	106			2–8%	6
		5.5–6.5%	4			3–5%	5
						3–8%	2
						10–15%	2
						< 1%	20
						< 4%	17
						< 5%	24
						< 7%	1
						< 8%	2
						< 10%	1
Total	72	**Total**	245	**Total**	**61**	**Total**	**108**

Table 4.
Percentage of carbon black.

3.3.2 Paraffin wax

Table 5 shows that paraffin wax content is in the 2–15% range in these brands. Paraffin wax is a key substance that lubricates other components to improve friction resistance. Brand B and Brand B's toner cartridges contain more uniform ranges. However, Brand D has various ranges, which are higher than the other half. For Brand A, a similar chemical type is wax, so paraffin is replaced by wax in toner cartridges.

3.3.3 Polyester resin

According to **Table 6**, all the four brands contain more than 50% polyester resin, the main chemical in toner. About 60% of the toner comprises polyester resin as it is used for melting the ingredients during printing. The data show that this percentage matches the market products. Some percentages are hidden by Brand B because the concentration of polyester resin may be an important survival skill and secret in the printer market.

3.3.4 Silica

Because functions of silica and iron oxide are similar, **Table 7** can be compared with **Table 8**. The range of silica in **Table 7** is approximately 1–3%. However, in **Table 8**, the range of iron oxide is approximately 40–50%. Iron oxide accounts for about 40% of the ingredients, matching what the data tell. It is worth mentioning that the less is the silica content, the more is the effect of iron oxide.

3.4 Discussion of function of main chemicals in toner cartridges

MSDS does not explain all chemicals. It only provides concentrations and percentages of chemicals in toner cartridges. According to MSDS, pigment, wax, silicon dioxide (silica), polymer, and water are the main chemicals in toner cartridges.

Brand A		Brand B		Brand C		Brand D	
Concentration	**Amount**	**Concentration**	**Amount**	**Concentration**	**Amount**	**Concentration**	**Amount**
		2–3%	10	< 2%	10	0.5–2%	3
		2–4%	18	< 5%	1	0.5–3%	1
		3–4%	4	< 8%	3	0.5–5%	1
		3–5%	4	< 10%	3	1–2.5%	2
		4–5%	4	< 15%	5	1–5%	22
						5–8%	12
						5–10%	41
						8–15%	5
						< 2%	23
						< 3%	12
Total	**0**	**Total**	**40**	**Total**	**12**	**Total**	**120**

Table 5.
Percentage of paraffin wax.

Brand A		Brand B		Brand C		Brand D	
Concentration	**Amount**	**Concentration**	**Amount**	**Concentration**	**Amount**	**Concentration**	**Amount**
5–10%	7	75–90%	21	0.2 < 1%	1	5.6%	4
45–55%	7	**	71	< 10%	19	55–95%	1
80–90%	2			< 50%	6	60–74%	5
85–95%	15			< 55%	20	75–95%	11
				< 65%	2	80–90%	4
				< 74%	5	80–95%	9
				< 85%	1	85–95%	8
						90–95%	5
						< 8.5%	8
						> 81%	2
						> 82%	1
						> 87%	13
						> 88%	29
Total	**31**	**Total**	**98**	**Total**	**54**	**Total**	**100**

****The exact percentage is hidden by Brand B.*

Table 6.
Percentage of polyester resin.

Brand A		Brand B		Brand C		Brand D	
Concentration	**Amount**	**Concentration**	**Amount**	**Concentration**	**Amount**	**Concentration**	**Amount**
< 2%	4			< 1%	10	0.2%	4
1–3%	126			< 2%	3	0.5–1%	3
				< 3%	3	0.5–2%	4
						1–2%	12
						1–3%	40
						1–4%	5
						1–5%	15
						1–11%	1
						2–4%	8
						< 2%	31
						< 3%	12
Total	**130**	**Total**	**0**	**Total**	**16**	**Total**	**135**

Table 7.
Percentage of silica in four brands.

3.4.1 Pigment

Pigments are one of the most popular colorants in toner and play a vital role in providing color on paper. Cyan, Magenta, Yellow, and Black are the four basic colors.

Brand A		Brand B		Brand C		Brand D	
Concentration	**Amount**	**Concentration**	**Amount**	**Concentration**	**Amount**	**Concentration**	**Amount**
30–40%	3			< 40%	7		
35–45%	3			< 45%	4		
40–50%	23			< 50%	41		
45–55%	1			< 55%	4		
Total	**30**	**Total**	**0**	**Total**	**56**	**Total**	**0**

Table 8.
Percentage of iron oxide.

Every color is a product of mixing of these four colors in different proportions. These four colors are used separately in the toner.

3.4.2 Wax

The data show that different types of waxes, such as paraffin, ester wax, and ethylene wax are used in toners. Toner contains wax additives and has physical properties such as friction resistance, solvent or grease corrosion resistance. In addition, the coefficient of friction is controlled to control the slip resistance. In other words, wax is a lubricating substance.

3.4.3 Silicon dioxide (Silica)

Silicon dioxide, also known as silica, is used to increase the chemical level of the charged portion. It develops the performance of ink absorption, drying speed, vivid reproduction of color, and so on. It functions like iron oxide. However, the amount of iron oxide reported was found to be too low, with only about 86 of the 1473 cartridges having it. Since iron oxide was an early mixture, silicon dioxide was developed to replace it. Therefore, silicon dioxide is one of the common materials in toner, not iron oxide.

3.4.4 Polymer

The polymer in the toner can melt and bond to the paper at high temperatures. Polyester resins, styrene acrylate copolymers, or styrene butadiene copolymer are examples of polymers used, depending on the manufacturer. However, there is a slight difference between styrene acrylic copolymers and polyester resins. Styrene acrylate copolymers have good adhesion to many surfaces, making them thermoplastic in nature, and polyester resins are used as adhesive resins in toners for high-speed printers.

3.4.5 Water

Water is a major component of Brand A cartridges and is related to the state of the toner. In general, water is good at dissolving ions and polar molecules, but is poor in dissolving non-polar molecules. Because of the polarity of its molecules, it interacts with charged polar matter differently from electrode-banded matter. Therefore, it acts as the primary medium and allows all other substances to dissolve each other. In addition, some dye particles can be dissolved in water, causing saturation color on the paper.

Table 9 summarizes and introduces the functions of main chemicals in different brands.

3.5 Health and environment impact of toner cartridge

Toner contains hazardous substances, which are flammable and toxic. They pose a huge risk to human health and environment [10–14]. Therefore, the toner industry is

	Chemicals	Function
Brand A	1,2-benzisothiazol-3(2H)-one	It is used as a preservative in products such as inks and photographic processing solutions.
	2-Pyrrolidinone	It is an intermediate in the manufacture of polymers and it can prevent the ink from evaporating into the air.
	Glycerin	It provides a good pigment dispersion stability and good wet-fastness and dry rub-fastness.
	Glycol	It improves the duration of time for which the substance mix stays mixed with water inside the ink cartridge.
	Water	It acts as the primary medium and allows dissolution of all other substances with each other.
Brand B	Carbon Black	It is the common name of black pigment and it appears black.
	Fatty Acid Ester	It improves the storage stability of the ink.
	PMMA	It acts as a plastic for melting in the printing process.
	Styrene-Acrylate Copolymer	It has good adhesion to many surfaces and makes it thermoplastic.
	Styrene-Acrylate Resin	It can impart good fixing and offset properties to provide good printing performance.
	Silicon Dioxide	It develops a strong performance of ink absorption, quick drying, vivid reproduction of color.
Brand C	Amorphous Silica	It develops strong performance of ink absorption, quick speed of drying, vivid reproduction of color.
	Pigment	It provides color onto the paper.
	Styrene Acrylate Copolymer	It has good adhesion to many surfaces and make it thermoplastic.
	Wax	It is for lubrication.
Brand D	Carbon Black	It is the common name of black pigment and appears black.
	Paraffin Wax	It is for lubrication.
	Polyester Resin	It is used as toner binder resin in high-speed printers.
	Silica	It develops a strong performance of ink absorption, quick drying, vivid reproduction of color.

Table 9.
Function of the main chemical in four brands.

supervised by the government and MSDS is prepared for each product. MSDS not only provides the information of chemicals and ingredients, but also prescribes some precautions.

These precautions are normally divided into five sections for safety regulations. They are "first aid measures," "firefighting measures," "accidental release measures," "handling and storage," and "exposure controls/personal protection." Understanding the precaution parts can help identify the health and environment impacts of toner cartridges because it involves outcomes of toner. Thus, "accidental release measures" and "handling and storage" are the parts for analysis of impacts.

Toner contains flammable and toxic harmful substances that pose great risk to human health and the environment. Therefore, all toner manufacturers are supervised by the government to prepare MSDS for each product. Understanding precautionary measures can help identify the health and environmental effects of toner cartridges, as they relate to toner results. Therefore, "accidental release measures" and "handling and storage" are the parts for analysis of impact.

3.5.1 Accidental release measures

"Personal precautions," "environmental precautions," and "methods and materials for containment and cleaning up" are the three parts in accident release measures. Description of precautions is relevant when the chemicals are released. **Table 10** shows a comparison between precautions suggested by the four brands, integrated by all MSDS. **Table 1** shows that Brand A has the largest number of chemical types, but its statement of precautions is the smallest. In other words, because of simple statements, Brand A has more potential influencing factors to humans and the environment. In addition, these chemicals mainly cause irritation in the context of personal health because they are also described as avoiding inhalation of dust. For the environment, these chemicals pollute water bodies, as companies warn people to keep chemicals away from water.

3.5.2 Handling and storage

The methods of storage depend upon properties of different chemicals. **Table 11** shows a comparison of the four brands in terms of processing and storage. In general, all the four brands have listed similar precautionary statements in MSDS. Prevention of generation of dust is the focus of general reminders because these powders irritate the body, especially the skin and eyes. These chemicals are kept away from direct sunlight, heat, and oxidants because they are flammable chemicals, and dangerous to humans and the environment. Therefore, safe storage of chemicals is an important process to keep chemicals safe.

3.5.3 Chemical safety issue

This instruction can help the user prevent unexpected release and storage problems from occurring in severe conditions. All statements of accidental release measures and storage in **Tables 10** and **11** are based on the characteristics of the chemicals. Therefore, these properties can be divided into different categories. There are seven categories (**Table 12**) that are used to integrate the two categories in **Tables 10** and **11** for simple interpretation based on local statutory regulation (factories and industrial undertakings (dangerous substances) regulations) [15].

Personal precautions	**Brand A**	• Avoid contact with skin, eyes and clothing.
	Brand B	• Avoid generation of dust
		• Do not breathe dust.
		• A suitable dust mask or dust respirator with filter type A/P may be appropriate
	Brand C	• Minimize dust generation and accumulation.
	Brand D	• Avoid inhalation of dust.
		• Wash thoroughly after handling.
		• Wear appropriate personal protective equipment. Ensure adequate ventilation.
Environmental Precautions	**Brand A**	• Keep out of waterways.
	Brand B	• Prevent substance entering sewers.
		• Washings must be prevented from entering surface water drains.
		• Prevent entry into waterways, sewers, basements or confined areas
	Brand C	• Do not flush into surface water or sanitary sewer system.
	Brand D	• Avoid dispersal of spilled material and contact with soil, ground and surface water, drains and sewers.
Methods and material for containment and cleaning up	**Brand A**	• Wipe up with adsorbent material (e.g. cloth, fleece).
	Brand B	• Sweep the spilt toner or remove it with a vacuum cleaner and transfer into sealed container carefully.
		• Sweep slowly to minimize generation of dust during cleanup.
		• If a vacuum cleaner is used, the motor must be rated as dust explosion proof.
		• Potential for very fine particles to be taken into the vacuum only to be passed back into the environment due to pore size in the bag or filter.
	Brand C	• Slowly vacuum or sweep the material into a bag or other sealed container. Clean remainder with a damp cloth or vacuum cleaner.
		• If a vacuum is used, the motor must be rated as dust explosion-proof.
		• Fine powder can form explosive dust-air mixtures.
		• Dispose of in compliance with federal, state, and local regulations
	Brand D	• Small spill: Remove source of ignition. Carefully wipe off with paper or wet cloth, avoiding inhalation of fine dust.
		• Large spill: Wear protective gear: respirator, rubber gloves, goggles. Do not use vacuum cleaner when a large amount is released.
		• This mixture like most finely divided organic powders, may create a dust explosion. Wipe up remainder with a wet cloth.

Table 10.
Accidental release measures.

Precautions for safe handling	**Brand A**	• Avoid contact with skin, eyes and clothing.
		• Clean contaminated surface thoroughly.
		• Ensure adequate ventilation.
	Brand B	• Keep out of the reach of children.
		• Avoid generation of dust.
		• Avoid inhalation of high concentration of dust.
		• Avoid contact with eyes.
	Brand C	• Keep out of the reach of children.
		• Avoid inhalation of dust and contact with skin and eyes.
		• Use with adequate ventilation.
		• Keep away from excessive heat, sparks, and open flames.
	Brand D	• Avoid breathing dust and contact with skin, eyes and clothing.
		• Handle in well ventilated work space.
		• Wash thoroughly after handling.
		• Treat from upwind position.
		• Keep away from excessive heat, spark, and open flame
Conditions for safe storage, including any incompatibilities	**Brand A**	• Keep in a dry, cool and well-ventilated place. Keep out of the reach of children.
		• Keep away from direct sunlight.
		• Keep away from heat and sources of ignition.
	Brand B	• Keep away from oxidizing agents.
		• Keep containers tightly closed in a dry, cool and well-ventilated place.
	Brand C	• Keep out of the reach of children.
		• Keep tightly closed and dry.
		• Store at room temperature.
		• Store away from strong oxidisers.
	Brand D	• Store in cool, dry and well-ventilated place.
		• Avoid direct sunlight.
		• Keep away from oxidizing materials.
		• Keep out of reach of children.

Table 11.
Handling and storage.

Categories	Statement
Corrosive	• Wear appropriate personal protective equipment. Ensure adequate ventilation.
Flammable	• Keep away from excessive heat, sparks, and open flames.
	• Keep away from direct sunlight.
	• If a vacuum cleaner is used, the motor must be rated as dust explosion proof.
	• Fine powder can form explosive dust-air mixtures.
	• Store at room temperature.
Harmful	• Avoid generation of dust
	• Minimize dust generation and accumulation.
	• Sweep the spilt toner or remove it with a vacuum cleaner and transfer into sealed container carefully.
	• Sweep slowly to minimize generation of dust during cleanup.
Irritant	• Avoid inhalation of dust. Wash thoroughly after handling.
	• Avoid inhalation of dust and contact with skin and eyes.
Organic substance	• Wipe up with adsorbent material (e.g. cloth, fleece).
Oxidizing	• Keep away from oxidizing agents
Polluted	• Keep out of waterways
	• Washings must be prevented from entering surface water drains.
	• Do not flush into surface water or sanitary sewer system. Avoid dispersal of spilled material and contact with soil, ground and surface water, drains and sewers.

Table 12.
Categories integrated by MSDS.

4. Conclusion

This study assembles, integrates, compares, and analyses data of 1473 toner cartridges of the four target brands. Most of the components in the target model are similar to search information. At the same time, the relationship between the percentage and function of major chemicals was studied. Finally, the health and environmental effects are discussed on the basis of safety measures suggested in MSDS of the four brands.

Brand A, Brand B, Brand C, and Brand D are well-known companies for professional products in the printing industry. Although raw materials used in toners are different, they maintain stable production for target customers. There are different chemicals, but most of them have the same function in the toner. For example, polyester resins, pigments, and waxes are essential chemicals in toner. The percentage in MSDS represents the weight of the chemicals in toner. The results show that some chemicals are changeable and can be combined with a variety of substances to remain competitive.

The disposal of waste toner has an absolute impact on human health and the environment. Basically, all MSDS contain some precautionary measures for helping customers avoid accidents and minimize the impact on environment. In addition, the manufacturers are aware of potential risks of toner cartridges and have provided some advice and recommendations to prevent hazards.

DOI: http://dx.doi.org/10.5772/intechopen.107439

5. Limitations

All the data in this study were from MSDS documents accompanying each toner model. The number and percentage of chemicals to analyze were based on MSDS. Since MSDS is published on the company's official website, the actual composition may differ from the documentation. Due to the lack of data of all toner models, concentration of each component in the toner could not be determined; it was not possible to purchase 1473 cartridges for analysis. Studying 1473 cartridges with real products is a time-consuming process involving comparison of large quantities of chemicals.

In addition, there are many brands of toners and this study focused on only four brands. For the MSDS, although documents provide specific names of chemicals and other useful information, the chemicals used for a color cannot be identified clearly. MSDS of four brands provide the color types in the list but not all the color types are the same. In other words, there is a lack of standard colors so that the colorant is unable to be compared and identified clearly.

6. Recommendations

In the sampling of specimens, more toner brands should be selected for analysis so that the analysis is more in-depth and comprehensive. To understand the properties of toner, properties of chemicals should be tested through physical and chemical testing. In addition, components of toner can be identified by infrared spectroscopy or chromatography. There are machines for identifying and analyzing compounds, so most of the chemicals in toner can be compared to MSDS. Today, printing technology is becoming more and more advanced in the global market. The performance of toner is the main research to improve the customer base to increase the company's business.

Acknowledgements

This study is part of a final year project submitted by K.S.W. Kwan in partial **fulfillment** of the requirements for her BA (Hons) degree in Fashion and Textiles in Institute of Textiles and Clothing, The Hong Kong Polytechnic University. This work is financially supported by The Hong Kong Polytechnic University (Account code: ZJM8).

Conflict of interest

The authors declare no conflict of interest.

Author details

Kelvin Sin-Wang Kwan and Chi-Wai Kan*
School of Fashion and Textiles, The Hong Kong Polytechnic University, Kowloon, Hong Kong

*Address all correspondence to: tccwk@polyu.edu.hk

References

[1] Pirela SV, Pyrgiotakis G, Bello D, Thomas T, Castranova V, Demokritou P. Development and characterization of an exposure platform suitable for physico-chemical, morphological and toxicological characterization of printer-emitted particles (PEPs). Inhalation Toxicology. 2014;**26**(7):400-408. DOI: 10.3109/08958378.2014.908987

[2] Bello D, Martin J, Santeufemio C, Sun Q, Bunker KL, Shafer M, et al. Physicochemical and morphological characterisation of nanoparticles from photocopiers: Implications for environmental health. Nanotoxicology. 2013;**7**(5):989-1003. DOI: 10.3109/17435390.2012.689883

[3] Pirela SV, Sotiriou GA, Bello D, Shafer M, Bunker KL, Castranova V, et al. Consumer exposures to laser printer-emitted engineered nanoparticles: A case study of life-cycle implications from nano-enabled products. Nanotoxicology. 2015;**9**(6):760-768. DOI: 10.3109/17435390.2014.976602

[4] Chalbot MCG, Pirela SV, Schifman L, Kasaraneni V, Oyanedel-Craver V, Bello D, et al. Synergistic effects of engineered nanoparticles and organics released from laser printers using nano-enabled toners: Potential health implications from exposures to the emitted organic aerosol. Environmental Science: Nano. 2017;**4**:2144-2156. DOI: 10.1039/C7EN00573C

[5] Morimoto Y, Ogami A, Kochi I, Uchiyama T, Ide R, Myojo T, et al. Continuing investigation of effect of toner and its by-product on human health and occupational health management of toner. Sangyo Eiseigaku Zasshi (Journal of Occupational Health). 2010;**52**(5):201-208. DOI: 10.1539/sangyoeisei.a10002

[6] Ari A. A comprehensive study on gas and particle emissions from laser printers: Chemical composition and health risk assessment. Atmospheric Pollution Research. 2020;**11**(2):269-282. DOI: 10.1016/j.apr.2019.10.013

[7] Karrasch S, Simon M, Herbig B, Langner J, Seeger S, Kronseder A, et al. Health effects of laser printer emissions: A controlled exposure study. Indoor Air. 2017;**27**:753-765. DOI: 10.1111/ina.12366

[8] International Agency for Research on Cancer. IARC Monographs on the Evaluation of Carcinogenic Risks to Humans, No. 65: Printing processes and printing inks, carbon black and some nitro compounds. Lyon, 1996

[9] Staista. Global printer market share by vendor 2015-2018. 2019. Available from: https://www.statista.com/statistics/541347/worldwide-printer-market-vendor-shares

[10] Martin J, Bello D, Bunker K, Shafer M, Christiani D, Woskie S, et al. Occupational exposure to nanoparticles at commercial photocopy centers. Journal of Hazardous Materials. 2015;**298**:351-360. DOI: 10.1016/j.jhazmat.2015.06.021

[11] Pirela SV, Martin J, Bello D, Demokritou P. Nanoparticle exposures from nano-enabled toner-based printing equipment and human health: State of science and future research needs. Critical Reviews in Toxicology. 2017;**47**(8):678-704. DOI: 10.1080/10408444.2017.1318354

[12] Salthammer T, Schripp T, Uhde E, Wensing M. Aerosols generated by

hardcopy devices and other electrical appliances. Environmental Pollution. 2012;**169**:167-174. DOI: 10.1016/j.envpol.2012.01.028

[13] Elango N, Kasi V, Vembhu B, Poornima JG. Chronic exposure to emissions from photocopiers in copy shops causes oxidative stress and systematic inflammation among photocopier operators in India. Environmental Health. 2013;**12**(1):78. DOI: 10.1186/1476-069X-12-78

[14] Yang CY, Huang YC. A cross-sectional study of respiratory and irritant health symptoms in photocopier workers in Taiwan. Journal of Toxicology and Environmental Health Part A. 2008; **71**(19):1314-1317. DOI: 10.1080/15287390802240785

[15] Labour Department. A guide to the factories and industrial undertakings (dangerous substances) regulations. Hong Kong. 2006

Chapter 5

Phenazines and Photoactive Formulations: Promising Photodrugs for Photodynamic Therapy

Ranulfo Combuca da Silva Junior,
Katieli da Silva Souza Campanholi,
Flávia Amanda Pedroso de Morais,
Laura Adriane de Moraes Pinto, Fabiana dos Santos Rando,
Magali Soares dos Santos Pozza and Wilker Caetano

Abstract

Photodynamic Therapy (PDT) is a therapeutic modality that can be applied with many photosensitizing compounds (PS). Photosensitization has shown promising results in damage against abnormal cell growth as cancer and inactivating a broad spectrum of microorganisms with no reported microbial resistance. Photodynamic processes occur by the light action at the appropriate wavelength in the presence of a PS that will be excited by the energy absorbed from the light source, where the interaction with the oxygen present in the cell will generate reactive oxygen species (ROS). The potential of phenazines as a photosensitizer is reviewed in this chapter as a practical guide to the future development of formulations that are effective for cancer treatment and microorganism control. Here we mainly summarize articles about phenazines from 2005 to 2021 when we performed a systematic search in the Science Direct, PubMed, Google Scholar, Web of Science, and Scopus databases. The carrier systems formed by micellar copolymers type Pluronic® have demonstrated effectiveness in incorporating several PS, ensuring its monomeric form for PDT applications. The fundamentals of the photosensitization mechanism are discussed. Studies have shown the beneficial impact of an appropriate incorporation technique to enhance the cellular uptake of phenazines compounds.

Keywords: phenazines, photodynamic therapy, photosensitizing agents, nanoplatform, micelles

1. Introduction

Success in treating diseases such as cancer or microbial diseases directly depends on the therapeutic modality applied. Given the side effects presented and the limitation in the efficiency of traditional procedures (surgery, chemotherapy, and radiotherapy), other alternatives are constantly being proposed in oncology [1]. Among the most promising modalities, Photodynamic Therapy (PDT) stands out as it does not present serious side effects and does not have limited efficiency [2]. PDT and antimicrobial Photodynamic Therapy (aPDT) are medical modality that has high specificity and selectivity in the treatment of infections caused by a virus, bacteria, protozoa, and fungi, as well as several cardiovascular, dermatological, and other diseases related to abnormal cell growth as cancer [2–5].

The PDT efficiency directly depends on a photosensitizing compound (PS) with ideal properties for the photophysical and photochemical processes that leads to the formation of singlet oxygen ($^1O_2$) and/or reactive oxygen species (ROS) that cause cell damage [2, 6]. PDT aims at the localized damage of living tissue with abnormal cell growth through its necrosis or infeasibility [7]. Likely targets of PDT are mitochondria, plasma membrane and other cell organelles, tumor cell nucleus, and blood vessels [8, 9]. This selectivity occurs due to the high concentration of lipoprotein receptors in neoplastic cells, where PS accumulates preferentially in diseased tissues, forming intravascular complexes with low-density proteins (LDL) [10].

The $^1O_2$ is a free radical produced during PDT and other related therapies. It is made when light activates the photosensitive compound administered to the patient. The $^1O_2$ action mechanism is based on its ability to cause damage to DNA, proteins, and other cellular molecules, leading to cell death [2]. Lipid peroxidation occurs when $^1O_2$ reacts with lipids in cell membranes, causing the formation of secondary free radicals and damaging the cell membrane structure. DNA damage can arise when $^1O_2$ reacts with DNA nucleotides, causing damage and interfering with replication and gene expression. The $^1O_2$ stops energy production when reacting with the enzymes of the respiratory cell chain and leads to cell death. The $^1O_2$ can react with cellular ribosomes and interrupt protein synthesis, causing cell death. The $^1O_2$ is an essential agent in photodynamic therapy and other radiation therapies, as these action mechanisms lead to the destruction of cancer cells and a reduction in the size of tumors [2].

Among the main characteristics of adequate PS is their low toxicity in the dark, light absorption between 400 and 850 nm (therapeutic window), high molar absorptivity values, and considerable formation of ROS inherent in the technique [11].

The development of promising drugs requires science to improve investigations based on the action of a compound or the joint effort of two or more drugs, thus synergy [12]. The first generation of phototherapeutic agents used in PDT is based on mixtures of porphyrin derivatives [7]. The search for PS with better optical and pharmacokinetic characteristics gave rise to the second generation of PS, similar to porphyrin molecules such as benzoporphyrins, chlorins, texapyrins, phthalocyanines, and naphthalocyanines [13]. Some of those PS compounds are already approved by the Food and Drug Administration (FDA/USA) for clinical applications using PDT. In addition, some countries are already approved, for example, Photofrin®, Levulan Kerasticks®, and Visudyne® (Verteporfin) [14].

Due to advances in research, studies, and treatments based on PDT, there is a great demand for new naturally occurring or synthetic PS that are biocompatible and have adequate properties [15, 16]. Over time, hematoporphyrin derivatives have been replaced by various PS compounds [17]. The third generation of PS is based on

DOI: http://dx.doi.org/10.5772/intechopen.110588

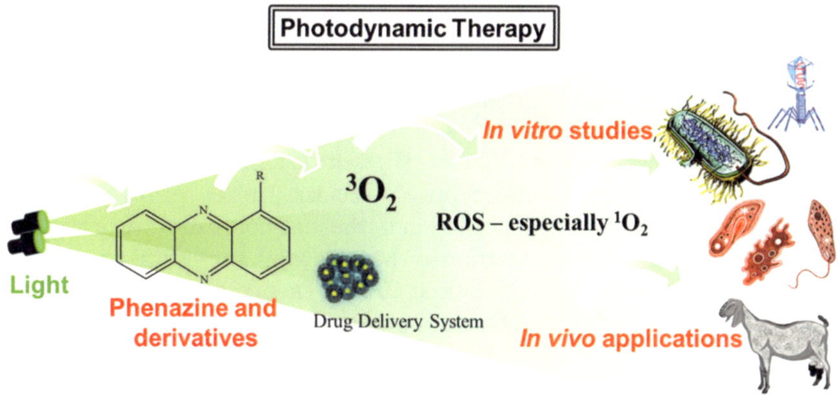

Figure 1.
Phenazines compounds incorporated into drug delivery system have great potential for photodynamic therapy applications.

formulations with different drug delivery systems (DDS) that provide better solubility in physiological media, more favorable pharmacokinetics, and allow the use of light sources with more remarkable penetration power into tissues.

Due to economic and environmental considerations, PS compounds obtained from abundant raw materials attract more interest than those prepared by complex chemical routes [15, 18]. Studies involving naphthodianthrones (hypericin), phenothiazines (methylene blue and toluidine blue), phthalocyanines, chlorins (chlorophyll A), xanthenes (rose bengal), and curcuminoids (curcumin) stand out in the literature [14]. These PS compounds are extensively studied *in vitro*, *in vivo*, and in photodiagnostic aspects [19–21].

This particular chapter presents explicitly an overview of compounds of the phenazine class (**Figure 1**), in which promising photoactive drugs such as neutral red (NR) [22], phenosafranin (PhS) [23], and safranine-O (Sf) [24], still poor explored for PDT applications.

2. Search strategy

The strategy we performed a systematic literature search in Science Direct, PubMed, Google Scholar, Web of Science, and Scopus databases using the combinations of the term "phenazines" with the following: "photodynamic therapy," "aPDT or antimicrobial Photodynamic Therapy, or PACT or photodynamic inactivation or PDI," "Animal studies or *in vitro* studies involving phenazines". Peer-reviewed articles published in English from 2005 to 2021 were included to compile this chapter. We have also scanned references for relevant articles.

3. Photosensitization mechanism

The combination of a PS compound with molecular oxygen ($^3O_2$) and visible light of the adequate wavelength generates ROS that causes cell components to oxidize and lead to death [2, 25]. In addition, PS reacts with neighboring molecules by electron or hydrogen transfer leading to the production of free radicals or by energy transfer to oxygen, inducing the production of $^1O_2$ [2, 26].

The photochemical processes originate from the interaction of light with matter which, by absorbing energy with adequate wavelength, allows the promotion of an electron from the ground state, called HOMO (Highest Occupied Molecular Orbital), to the excited state LUMO (Lowest Unoccupied Molecular Orbital), of higher energy [27]. The excited state is unstable, and in it, the molecule can suffer chemical processes (rearrangements or fragmentation of the molecule) or physical processes (deexcitation) [27]. According to the Molecular Orbital Theory, oxygen in the ground state and excited can assume different forms of occupation of molecular orbitals anti-binders, as shown in **Figure 2**.

In the ground state, molecular oxygen has two unpaired electrons in the doubly degenerate antibonding orbitals, π^*_x, and π^*_y. These electrons have the same spin, resulting in a maximum multiplicity and, thus, the lowest state oxygen energy. Therefore, the ground state of molecular oxygen is a triplet, which has the spectroscopic term $^3\Sigma_g$. The excited state of oxygen that has all valence electrons paired is singlet oxygen. Singlet oxygen has two forms with distinct symmetries, one of smaller energy $^1\Delta_g$, doubly degenerate ($^1\Delta_x$ and $^1\Delta_y$; 92.4 kJ mol^{-1}), and another one with higher energy ($^1\Sigma$; 159.6 kJ mol^{-1}). The second excited state of oxygen has a short lifetime since the transition to the $^1\Delta_g$ state is allowed by spin. The different symmetry of the $^1\Delta_g$ species with respect to the ground state and the spin prohibition of the $^1\Delta_g - {}^3\Sigma_g$ transition ensures that the $^1\Delta_g$ species has a long enough lifetime to allow oxidation of organic molecules [7].

The photochemical processes that occur in the excited state of photoactive molecules can be represented by the Jablonski diagram (**Figure 3**) [27, 28]. Among these processes, internal conversion (IC), fluorescence, intersystem crossing (ISC), and phosphorescence stand out [27, 29].

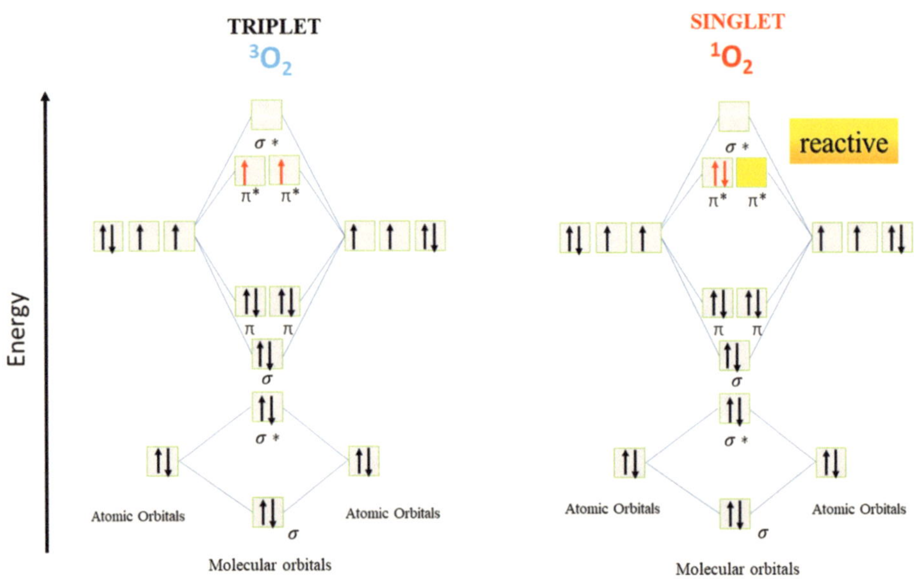

Figure 2.
Electronic distribution in the antibonding molecular orbitals for the oxygen electronic states. It was adapted from Lakowicz [27].

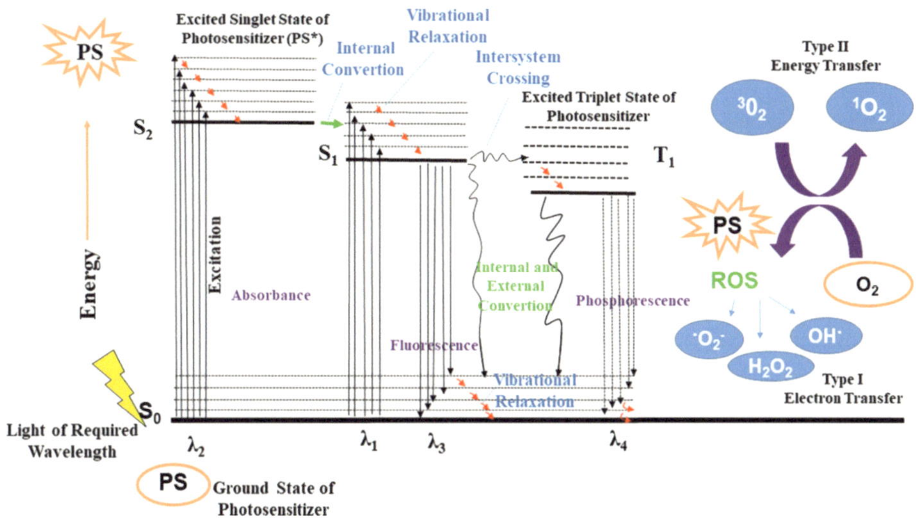

Figure 3.
Jablonski diagram: Electron transfer scheme by type I and type II mechanisms with singlet oxygen, superoxide anions, and hydroxyls production. It was adapted from Lakowicz [27].

After molecule excitation to the second excited state, the relaxation process for the first excited state can occur by IC (where the internal energy is lost by collision) and then return to the ground state via fluorescence emission. Deactivation can also arise by ISC between vibrational levels of the same energy and electronic states of different multiplicities. The decay from the triplet state to the singlet fundamental can then occur by phosphorescence emission or collisions [27, 29]. For PDT, it is interesting that the PS is preferably in the excited triplet state, as it will have the same spin multiplicity of $^3O_2$ favoring the photochemical processes of producing $^1O_2$ and ROS. The interaction with biological substrates in PDT can occur by two mechanisms: Type I, where photo-oxidation occurs through the transfer of electrons between the triplet state of PS and the substrate forming radical ions that react with oxygen in the ground state resulting in ROS. The Type II mechanism involves the energy transfer from the triplet excited state of PS to molecular oxygen generating $^1O_2$ [2, 28].

The literature reports the use of PDT in treating primary carcinomas or metastases in the head and neck regions, such as the oral cavity, pharynx, and larynx [30]. However, after tumor recession, secondary diseases such as those caused by bacteria may manifest. Photodynamic treatment circumvents this problem as PDT is not restricted to microorganisms.

With the increasing cases of acquired resistance by bacteria against antibiotics, the search for the control of microorganisms via PDT also attracts interest from the scientific community [31]. As a result, PDT has been widely applied in the microbiological area, standing out as a promising technique in microorganism inactivation [32]. Furthermore, PDT offers advantages over the usual antimicrobial agents, triggering rapid cell death and unlikely development of resistance by the microorganism [33].

4. Photosensitizer formulation

Several PS can be used for different applications because of their particular properties. Therefore, it is necessary to know the essential characteristics of each PS to use its therapeutic properties better. The most cited PS compounds in the literature present, for the most part, structures containing aromatic rings or conjugated systems, which give them high hydrophobicity. In this sense, efforts in developing strategies in the formulation of PS administration stand out. However, even PS with a small structure presents a self-aggregation problem, making it challenging to apply in aqueous media.

Incorporating PS compounds into nanostructured systems aims to minimize the effects of self-aggregation in aqueous media, protect PS against degradation and elimination by the organism, and facilitate the biotransport. In some cases, the adequately incorporated PS assists the vectorization of the nanostructured system, increasing the bioavailability of the photoactive drug at the application site (third generation), promoting a controlled release in the regions (tissues and organs) to be treated [34]. In addition, they decrease side effects (toxicity) and microbial resistance. In this sense, several carrier systems are generally constituted by colloidal dispersion systems, such as copolymeric micelles [35], liposomal vesicles [36], dendrimers [37], cyclodextrins [38, 39], polymer-DNA complexes (polyplexes) [40], nanogels [41], nanotubes [42], nanosuspensions [43], nanocrystals [44], solid lipid nanoparticles [45], metallic or ceramic nanoparticles [46], among other nanoscale materials for medical use are being widely studied [47–49]. **Figure 4** illustrates the most commonly found nanoparticulate carrier systems in the literature for carrying PS [50].

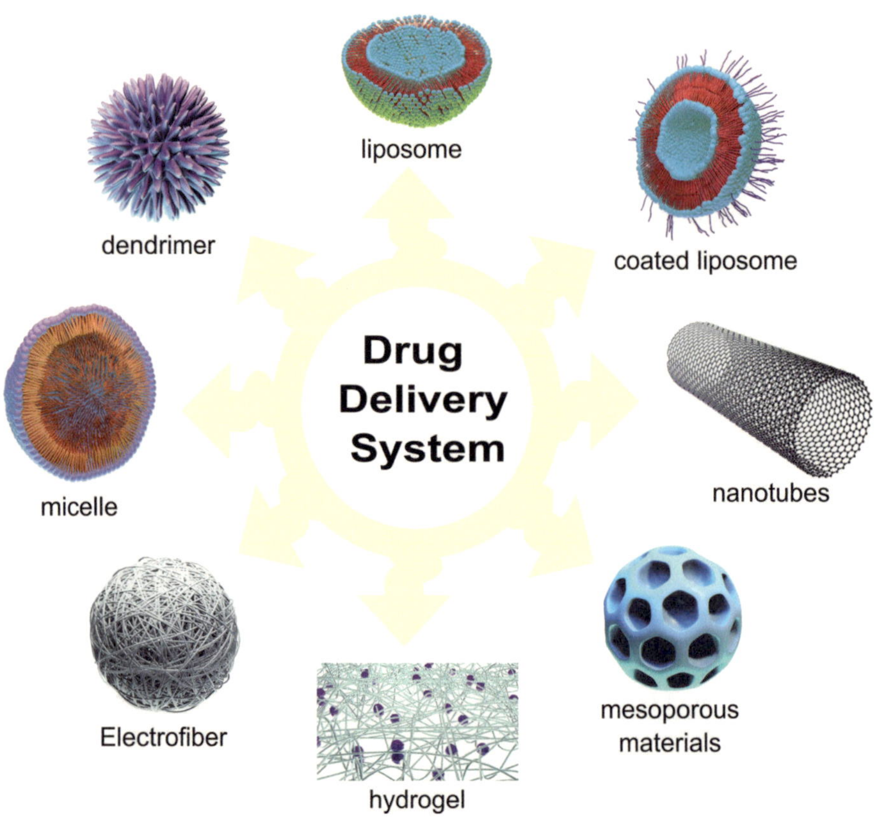

Figure 4.
Exemplification of some carrier systems currently used. I was adapted from Senapati et al. [50].

All these carrier systems have the advantage of protecting PS as a common characteristic. In addition, they contribute to keeping the PS monomeric form ideal for application in PDT [48, 51]. The leading site of photodynamic action is the cell membranes; therefore, it is of great importance that the carrier system is similar or presents a specific interaction with it. The PS efficiency in PDT applications depends on the microenvironment in which the PS is located so that there are no changes in its physicochemical and photophysical properties and in its interaction with living biological systems [36]. Parameters such as bioaccessibility, passive transport, and permeation of the drug into the membrane are of great importance in developing new drugs for PDT applications [11, 52]. Micellar and liposomal environments, for example, are reasonable models of cellular environments and provide useful information about PS molecules against organized systems and biomimetics [5, 41, 52, 53].

Since the cell membrane is a complex system composed of a lipid bilayer consisting of several types of phospholipids, cholesterol, glycolipids, and proteins, the copolymer micelles are good carrier systems for biomimetic the cell membrane [54]. Different micellar microdomains, called hydrophobic and hydrophilic, allow for estimating the drug partition tendency according to the region where it was solubilized. Compared to membrane models, polymeric micelles have the following advantages: minimal toxicity, narrow size distribution, longer residence time in the circulatory system, improved bioavailability, and more excellent stability of the incorporated photosensitizer [54].

5. Polymeric nanoparticles: copolymeric micellar systems in drug formulation

Most dyes, including phenazines, are characterized by forming aggregates in an aqueous solution harming their application. For this case, using DDS systems such as copolymeric micelles contributes to stabilization and solubilization, which maintains the PS characteristics inherent in PDT application [55]. In addition, these compounds are readily adsorbed by anionic micelles due to their distinct and amphiphilic character [56]. Therefore, studying the interaction of dyes with micellar systems is essential for biological applications since PS-associated biopolymers provide high drug efficacy with reduced toxicity [57].

Polymeric nanoparticles have been extensively applied in the pharmaceutical industry for drug formulation [58–60]. Currently, research is directed toward studying the association of photoactive molecules with nanostructured systems aiming for a specific carrier for controlled release in the regions (tissues and organs) to be treated [58]. Colloidal copolymers stand out, forming micelles (triblock systems) and effectively stabilizing hydrophobic molecules in aqueous media, such as PS, that remain in the form of monomers [16].

Polymeric colloidal systems are excellent for drug delivery and have been widely explained in several studies [48, 53]. These carrier systems can be classified as anionic, cationic, zwitterionic, and nonionic (according to their state of charge) [6]. Micelles are amphiphilic structures composed of a hydrophobic and a hydrophilic region (**Figure 5A**) [52]. In the aqueous phase, the micelles keep the hydrophobic portion facing the inside of the structure (lipophilic core) [61].

Micelles simulate the interface of a biomembrane and provide a biomimetic environment for the study of specific interactions, as well as the penetration and location of PS in the intracellular domain [4, 5]. Furthermore, studies have shown that PS

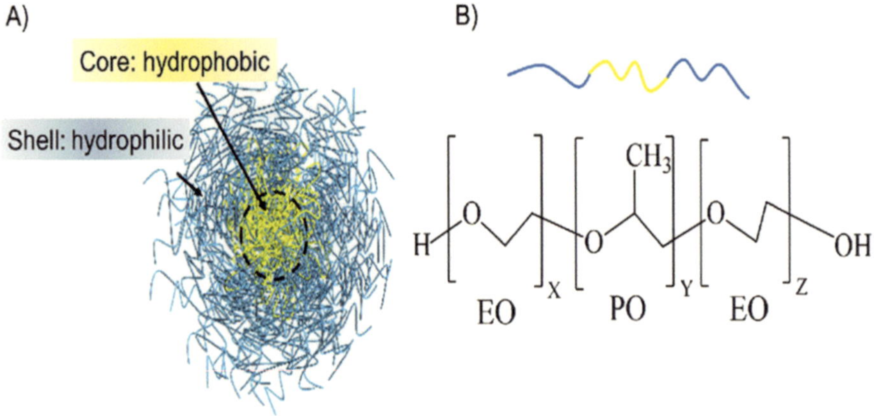

Figure 5.
(A) General monomer structure of polymeric micelle, and (B) general structure of the copolymeric monomer (Pluronic® class).

incorporation into micellar systems can reduce toxicity, target PS to specific sites and improve permeation and bioavailability in topical applications [11, 41]. Given these particularities, the biotransport and delivery of the drug in the target tissue become a particular challenge in PDT [4].

Several studies point to the use of surfactants of the Pluronic® class (or Poloxamers) [62]. The copolymer unimers (**Figure 5B**) consist of triblock molecules of repeating units of oxyethylene (EO – hydrophilic) and oxypropylene (PO – hydrophobic) with the following configuration: $(EO)_X(PO)_Y(EO)_X$ [63]. This group of copolymers is non-toxic, biocompatible, and has binding sites suitable for the solubilization of hydrophobic drugs [63]. In addition, triblock copolymers have a low critical micellar concentration (CMC, $10^{-6} - 10^{-7}$ mol L^{-1}) and the micelles formed to have high thermodynamic and kinetic stability, which guarantees a slow destructuring for unimers when they are exposed to an environment where the concentration is below the CMC [64, 65]. Another factor to be considered for studies with polymeric surfactants is the temperature. The micellization process will only occur above a specific critical micellar temperature (CMT), which, in turn, is a function of the surfactant concentration [64, 65]. In addition, there is evidence that temperature increases also cause an increase in the aggregation number (N_{ag}) of the copolymers [64]. Therefore, to be effective, the polymer used to obtain the micelle must present CMT compatible with the conditions necessary for the application [65].

Due to their favorable properties, the triblock copolymers of the Pluronic® class are potentially useful for drug delivery and controlled release systems [4, 53]. Furthermore, compared to other membrane models, copolymeric micelles have the following advantages: minimal toxicity, narrow size distribution, longer residence time at the application site, improved bioavailability, and more excellent stability of the incorporated PS [54]. These characteristics allow us to describe the micellar copolymer system as promising for clinical applications [66, 67].

6. Phenazines compound as photosensitizers or photoactive drugs

Phenazines ($C_{12}H_8N_2$) are heterocyclic aromatic nitrogenous compounds and the most important backbone is a pyrazine ring (1,4-diazobenzene) with two annulated benzenes (**Figure 6**) [68].

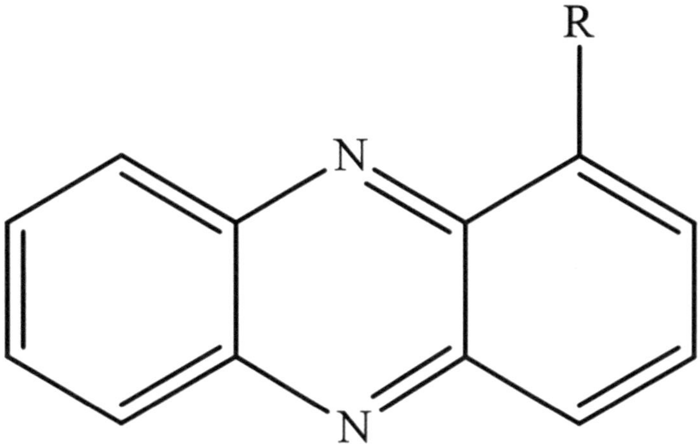

Figure 6.
General chemical structure of phenazines compounds.

This compound class is classified as redox-active secondary metabolites and pigmented. They are produced by fluorescent *Pseudomonas* species and have antimicrobial, antiparasitic, neuroprotective, insecticidal, anti-inflammatory, and anticancer activity [69].

Phenazines can undergo redox cycling in the presence of various reducing agents and molecular oxygen, which leads to the accumulation of superoxide (O_2^-) and hydrogen peroxide (H_2O_2), causing oxidative cell injury or death [68]. This property can be better explored considering the photochemical/photophysical potential of phenazines in PDT applications.

Natural phenazines are isolated from marine and terrestrial microorganisms, including (*Pseudomonas ssp*., *Streptomyces ssp*., and *Actinomycete ssp*.), which generates an infinity of synthetic derivatives [69]. *Pseudomonas aeruginosa* is a Gram-negative bacterium most studied for its ability to produce phenazine-active pigments, such as pyocyanin, phenazine-1 carboxyamide, and pyorubrins [68]. Furthermore, *P. aeruginosa* can survive in varied environments such as soil and water and colonize plant and animal tissues [70].

The literature describes the antioxidant, anti-inflammatory, and non-cytotoxic properties of the red and yellow phenazynic pigments produced by *P. aeruginosa*. These pigments confer great potential for application in the pharmaceutical and cosmetic industry [71, 72]. Phenazine derivatives differ in their chemical and physical properties based on the type of functional group position present, and they can also be used against the proliferation of various cancer cell lines [73–76]. There are more than 100 different compounds of natural origin and over 6000 synthetic compounds. Many of these compounds have been investigated as potential anticancer agents and microorganism control [77]. Some *in vitro* studies using phenazine derivatives are shown in **Table 1**.

Most studies involving compounds of the phenazines class have been reported promisingly in the dark [23]; however, the photodynamic potential cannot be disregarded. The literature presents scattered articles concerning different studies of phenazines derivative compounds performed *in vitro*, but without continuity or emphasis on the photophysical potential of phenazines, as well as *in vitro* and *in vivo* studies. Further, this group of compounds has also been used to develop color-emitting materials [91] or fluorescent biosensors [92].

Phenazines derivatives	Obtaining	Applications	Reference
2-chloroacetylamino-7(8)-nitrophenazine N5,N10-dioxide 2-amino-7(8)-(1,3-dioxol-2-yl)phenazine N5,N10-dioxide(2),2-chloroacetylamino-7(8)-(1,3-dioxol-2-yl)phenazine N5,N10-dioxide 2-amino-7(8)-methoxyphenazine N5,N10-dioxide	Cytotoxins of 2-amino-or2-hydroxy phenazine5,10-dioxide derivatives	*in vitro* antitumoral effect against Caco-2 cells	[78]
phenazine 5,10-dioxide derivatives	Derivative from N-oxides containing heterocycles	*in vitro* growth inhibitors of *T. cruzi*	[79]
phenazine-1-carboxamide	Obtained from the *Pseudomona aeruginosa*	antibacterial activity of a *Pseudomonas aeruginosa*-derived compound against methicillin-resistant *S. aureus* (MRSA) strains	[80]
phenazin-1-ol phenazine-1-carboxylic acid 2-heptyl-3-hydroxyl-4(1H)-quinolone phenazine-1-carboxamide	Producing from *Pseudomona aeruginosa* fluorescent	Medically important fungi: *Aspergillus flavus* MTCC 183, *Candida albicans* MTCC 277, *Candida tropicalis* MTCC 184, *Cryptococcus gastricus* MTCC 1715, and *Trichophyton rubrum* MTCC 296. Agriculturally important fungi: *Fusarium oxysporum* MTCC 284, *Rhizoctonia solani* MTCC 4634, and *Penicillium expansum* MTCC 2006	[81]
phenazine 1-carboxamide	Produced by *Pseudomonas* strain MCC2142	fungi such as *Candida albicans*, *Candida glabrata*, *Cryptococcus neoformans*, *Fusarium oxysporum*, *Aspergillus fumigatus*, *Aspergillus niger* and *Benjaminiella poitrasii*	[82]
Pontemazines A and B	Isolated from the culture broth of *Streptomyces sp.* UT1123	neuronal cell protective effect on glutamate-induced mouse hippocampal HT-22 cell damage	[83]
A series of 2,3,7-trisubstituted phenazines	Introduction of carboxylic or carboxamide group in 2,3-dialkoxy-phenazine	*in vitro* on human pancreatic (MiaPaCa-2) cell lines	[74]
imidazo[4,5-b]phenazine-2-thione methylthio ethyl 1-aryl-3H-[1, 2, 4]triazolo[2,3-a]imidazo[4,5-b] phenazines ethyl (2Z)-3-aminophenazin-2-yl)amino] (phenylhydrazono) ethanoate pyrazino[2,3-b]phenazine [1, 4]diazepino[2,3-b]phenazine 2,3-dibenzoylaminophenazine 1H-Imidazo[4,5-b]phenazine 4-[(E)-(3-amino phenazin-2-yl)diazenyl]	2,3-Diaminophenazine was used as a precursor	all compounds were tested as inhibitors of the proliferation of human lung carcinoma and colorectal cancer cell lines through inhibition of Tyrosine Kinases	[73]

Phenazines derivatives	Obtaining	Applications	Reference
benzo[a]phenazin derivatives	Treatment of 2-fluoro-nitrobenzene or 2-chloro-3-nitropyridine	four human cancer cell lines (HL-60, K-562, HeLa, and A549)	[84]
16-(4-ethoxyphenyl)-3,3-dimethyl-2,3,4,16-tetrahydro-1H-benzo[a]chromeno[2,3-c]phenazin-1-one 16-(4-Ethoxyphenyl)-3,3-dimethyl-2,3,4,16-tetrahydro-1H-benzo[a]chromeno[2,3-c]phenazin-1-one 16-(2,5-Dimethylphenyl)-3,3-dimethyl-2,3,4,16-tetrahydro-1H-benzo[a]chromeno[2,3-c]phenazin-1-one 16-(Benzo[d][1, 3]dioxol-5-yl)-3,3-dimethyl-2,3,4,16-tetrahydro-1H-benzo[a]chromeno[2,3-c]phenazin-1-one 3,3-Dimethyl-16-(o-tolyl)-2,3,4,16-tetrahydro-1H-benzo-[a]chromeno[2,3-c]phenazin-1-one 16-(3-Bromophenyl)-3,3-dimethyl-2,3,4,16-tetrahydro-1H-benzo[a]chromeno[2,3-c]phenazin-1-one 16-(2-Bromophenyl)-3,3-dimethyl-2,3,4,16-tetrahydro-1H-benzo[a]chromeno[2,3-c]phenazin-1-one 16-(3-Hydroxyphenyl)-3,3-dimethyl-2,3,4,16-tetrahydro-1H-benzo[a]chromeno[2,3-]phenazin-1-one 16-(3-Methoxyphenyl)-3,3-dimethyl-2,3,4,16-tetrahydro-1H-benzo[a]chromeno[2,3-]phenazin-1-one 4-(3,3-Dimethyl-1-oxo-2,3,4,16-tetrahydro-1H-benzo-[a]chromeno[2,3-c]phenazin-16-yl)benzonitrile 16-(4-Fluorophenyl)-3,3-dimethyl-2,3,4,16-tetrahydro-1H-benzo[a]chromeno[2,3-c]phenazin-1-one 16-(2-Methoxyphenyl)-3,3-dimethyl-2,3,4,16-tetrahydro-1H-benzo[a]chromeno[2,3-]phenazin-1-one 16-(1H-Indol-3-yl)-3,3-dimethyl-2,3,4,16-tetrahydro-1H-benzo[a]chromeno[2,3-c]phenazin-1-one 3,3-Dimethyl-16-(3,4,5-trimethoxyphenyl)-2,3,4,16-tetrahydro-1H-benzo[a]chromeno[2,3-c]phenazin-1-one 3,3-Dimethyl-16-(thiophen-2-yl)-2,3,4,16-tetrahydro-1H-benzo[a]chromeno[2,3-c]phenazin-1-one 3,3-Dimethyl-16-(3-nitrophenyl)-2,3,4,16-tetrahydro-1H-benzo[a]chromeno[2,3-c]phenazin-1-one 16-(3-Fluorophenyl)-3,3-dimethyl-2,3,4,16-tetrahydro-1H-benzo[a]chromeno[2,3-c]phenazin-1-one 3,3-Dimethyl-16-(4-nitrophenyl)-2,3,4,16-tetrahydro-1Hbenzo[a]chromeno[2,3-c]phenazin-1-one 16-(4-(Dimethylamino)phenyl)-3,3-dimethyl-2,3,4,16-tetrahydro-1H-benzo[a]chromeno[2,3-c]phenazin-1-one 16-(4-Ethoxyphenyl)-2,3,4,16-tetrahydro-1H-benzo[a]-chromeno[2,3-c]phenazin-1-one16-(4-Isopropylphenyl)-2,3,4,16-tetrahydro-1H-benzo-[a]chromeno[2,3-c]phenazin-1-one 16-(2-Chlorophenyl)-2,3,4,16-tetrahydro-1H-benzo[a]-chromeno[2,3-c]phenazin-1-one 16-(2-Methoxyphenyl)-2,3,4,16-tetrahydro-1H-benzo-[a]chromeno[2,3-c]phenazin-1-one 16-(3-Methoxyphenyl)-2,3,4,16-tetrahydro-1H-benzo-[a]chromeno[2,3-c]phenazin-1-one 16-(3-Chlorophenyl)-2,3,4,16-tetrahydro-1H-benzo-[a]chromeno[2,3-c]phenazin-1-one 16-(3-Fluorophenyl)-2,3,4,16-tetrahydro-1H-benzo-[a]chromeno[2,3-c]phenazin-1-one	Synthesis by condensation reaction of benzo[a]phenazine-5-ol	antioxidant and anticancer activities against HeLa and SK-BR-3 cell lines	[75]
Phenazine-1-carboxamide	Isolated from *Pseudomonas sp.* strain PUP6	cytotoxic activity against lung (A549) and breast (MDA-MB-231) cancer cell lines	[85]

Phenazines derivatives	Obtaining	Applications	Reference
phenazine-1,6-dicarboxylic acid phenazine-1-carboxylic acid	Produced by *Lactococcus* BSN307 strain	antifungal activity against *Aspergillus niger*, *Penicillium chrysogenum* as well as *Fusarium oxysporum* cytotoxicity against cancer cell lines HeLa and MCF-7 and normal H9c2 cells	[86]
phenazine-1-carboxamide	Isolated from the bacterium *Pantoea agglomerans* naturally present in soil	cytotoxicity on the cancer cell lines A549, HeLa, and SW480	[87]
Pyrano[3,2-a]phenazine derivatives: 3-amino-10-((1-aryl-1H-1,2,3-triazol-5-yl) methyl)-20-oxospiro[benzo[a] pyrano[2,3-] phenazine-1,30-indoline]-2-carbonitrile	Synthesized from 2-amino-3-hydroxyphenazine	cytotoxicity against HCT116, MCF7, HepG2 and A549 cancer cell lines *in vitro*	[76]
Riminophenazine	Synthesized from clofazimine and TMP-phenazines	*in vitro* antiplasmodial, cytotoxic, and oxidative activities against HeLa cells	[88]
N-((tert-Butyldimethylsilyl) oxy)-9-chlorophenazine-1- carboxamide 9-Chloro-N-hydroxyphenazine-1-carboxamide 9-Chlorophenazine-1-carboxamide 9-Chloro-N-cyanophenazine-1-carboxamide 9-Chloro-N-(1H-tetrazol-5-yl) phenazine-1-carboxamide 9-Chlorophenazine-1-carbonitrile 2-((2,5-Dichlorophenyl)amino)-3-nitrobenzoic acid 6,9-Dichlorophenazine-1-carboxylic acid Methyl 6,9-dichlorophenazine-1-carboxylate 6,9-Dichlorophenazine-1-carboxamide 6,9-Dichloro-N-(methylsulfonyl) phenazine-1-carboxamide 9-Bromo-6-methoxyphenazine-1-carboxylic acid 7-Bromo-9-methoxyphenazine-1-carboxylic acid 6,9-Dimethoxyphenazine-1-carboxylic acid 9-Bromo-6-methoxy-N-(methylsulfonyl) phenazine-1-carboxamide 9-Bromo-N-cyano-6-methoxyphenazine-1-carboxamide Methyl 7-bromo-9-methoxyphenazine-1-carboxylate 9-Bromo-7-methoxyphenazine-1-carboxylic acid (25) and 7-methoxyphenazine-1-carboxylic acid Methyl 9-bromo-7-methoxyphenazine-1-carboxylate Methyl 7-methoxyphenazine-1-carboxylate N-(Methylsulfonyl)phenazine-1-carboxamide N-Methyl-N-(methylsulfonyl) phenazine-1-carboxamide N-Cyano-9-fluorophenazine-1-carboxamide 9-Fluoro-N-(methylsulfonyl) phenazine-1-carboxamide 9-Methyl-N-(methylsulfonyl)phenazine-1-carboxamide Methyl 9-methoxyphenazine-1-carboxylate 9-Chloro-N-methyl-N-(methylsulfonyl) phenazine-1-carboxamide	Chloro-substituted phenazines containing acid bioisosteres	*in vitro* antimicrobial activity against Gram-positive (methicillin-resistant *Staphylococcus aureus*, MRSA) and Gram-negative (*Escherichia coli*) bacteria	[89]

Phenazines derivatives	Obtaining	Applications	Reference
phenazine analogue (CPUL1)	Synthesized from 2-amino-3-hydroxyphenazine	antitumor activities in initial stage of Hep G2 cells	[90]

Table 1.
Studies of phenazines derivatives compounds.

Udumula *et al.* [93] described the synthesis and biological evaluation of two phenazines natural products and a series of phenazines that show promising activities against methicillin-resistant *Staphylococcus aureus* associated with the CA-MRSA community with low minimum inhibitory concentration (MIC) values in the micromolar range [93]. The most active compound also showed good IC_{50} values against Human Keratinocyte Cell (HaCat) [93]. In work proposed by Krishnaiah *et al.* [89], the authors synthesized a series of phenazines, and the *in vitro* antimicrobial activity was evaluated against Gram-positive bacteria (methicillin-resistant *Staphylococcus aureus*, MRSA) and Gram-negative bacteria (*Escherichia coli*) [89]. These studies have indicated that the molecules do not disrupt bacterial membranes, and activity is not directly linked to the reactive oxygen species generation [89]. Therefore, despite a large number of studies on this compound class, this review highlights the great photodynamic potential of compounds of this class that can be used in the control of microorganisms and cancer treatment [94].

Neutral Red (NR) cationic dyes have great potential as photosensitizers. The exciting characteristic of NR as a PS is its water solubility compared with the majority of photosensitive and absorption bands located in the range of 550–700 nm [22]. Concerning with few phenazines studies as photodrugs explored in the literature (**Table 2**), for instance, the NR phenazines compound, when incorporated into Gold Nanoparticles, contributed to the *in vitro* tumor cell lines reduction [22], or when in aqueous medium significantly reduced *in vitro S. aureus* colonies in photoinactivation assays [98, 103].

Phenosafranin (PhS), another phenazines class compound, has the potential for use as a PS in PDT. It is well known that PhS absorbs strongly at 503–530 nm. Additionally, this dye presents a good photodynamic effect, namely low toxicity in the dark and higher toxicity under optical excitation [23]. PhS was incorporated into single-wall carbon nanotubes and showed satisfactory photodynamic effects against the BHK-21 cell line [42]. Furthermore, when encapsulated in liposomes (DMPC), PhS exhibited excellent *in vitro* activity against Human Cervical Carcinoma (HeLa) [23]. Another application for PhS found in the literature is *in vitro* photodynamic inactivation of *S. aureus* and *E. coli*. The bacterial cells were eradicated by ROS produced upon irradiation [102].

Sf, phenazine compound can also be used as a sensitizer in electron transfer reactions in a homogeneous medium [104–106], semiconductors [106], polymeric medium [104] and as probes in the reverse micellar system [107, 108] due to its absorbs band in 500–550 nm [56, 109]. In addition, the antimicrobial photodynamic ability of Sf is exploited for the inactivation of microorganisms, such as *Staphylococcus aureus* and *Escherichia coli* [24, 43], *Shigella flexneri*, *Bacillus subtilis* [24], periodontopathogenic bacteria (biofilms in periodontal treatment) [97], and mitochondrial oxidation [95].

Among the compounds of this class, the most cited for PDT applications is Sf, which presents some photophysical studies with properties already defined. Sf has properties that allow its use as a PS in PDT, such as considerable singlet oxygen quantum yield ($\Phi_\Delta{}^1O_2$) [33], an amphiphilic character that confers more significant interaction with biological substrates, does not present toxicity to healthy cells and generates reactive oxygen species (ROS), which inactivate microorganisms [101].

Although Sf presents several favorable properties to be applied in PDT, it has been poorly explored for this purpose. For example, spectroscopic studies on the interaction of Sf in liposomes L-egg lecithin phosphatidylcholine (PC) showed interaction. Still, the photodynamic effects of this formulation were not explored [57].

Sf incorporated into a silica matrix as a heterogeneous delivery system indicated that Sf has the potential for oxidation with singlet oxygen production [110]. Recent studies report Sf is incorporated into copolymer micelles [101]. This formulation ensured the Sf monomerization and preserved its physicochemical and photodynamic properties [33]. The use of triblock copolymers in the Sf incorporation guaranteed better results in PDT applications when compared to aqueous solutions [33].

Some authors evaluated the bactericidal effect of a serum combined with the action of Sf [24]. The authors did not use carrier systems, considering only the implications observed for the PS, not taking into account the self-aggregation processes characteristic of a PS in an aqueous medium. The photoefficacy of Sf in aqueous media was evaluated as an acaricide against female *Hyalomma dromedarii* ticks using *in vitro* immersion bioassays [99].

Similarly, Sasnauskiene *et al.* [95] evaluated the stimulated production of ROS by Sf in the inner space of mitochondria. The authors do not consider the self-aggregating effects resulting from the gradual release of PS in an aqueous medium [95].

Li *et al.* [96] explored the damage of bovine serum albumin (BSA) caused by Sf under ultrasonic irradiation [96]. The authors used Sf in an aqueous medium, a condition in which Sf tends to form self-aggregates that drastically reduce the $^1O_2$ formation. Therefore, we propose the use of low-cost biocarrier systems, Pluronic® class. This triblock copolymer was used to solubilize Sf and maintain its photophysical properties [33]. Such attributes favored the prevention and treatment of mastitis via PDT [33, 100, 101].

Table 2 presents some photochemotherapeutic antimicrobial studies of phenazines found in the literature.

Compound	Delivery system	Application	Illumination condition	Effect	Reference
Sf	Aqueous solution (1 mMdm^{-3})	Bactericidal effect of serum combined with the action of PS	White non mutagenic light (5Wcm^{-2})/90 min	The strains are sensitive to the photodynamic action	[24]
Sf	Aqueous solution (in PBS 0.7 μg mL^{-1})	Stimulated production of ROS in the inner space of mitochondria	LED at $\lambda = 509 \pm 5$ nm (29 Wm^{-2})/0.5–15min	Damage induced apoptosis	[95]
PhS	Single-wall carbon nanotubes modified	BHK-21 cell line (from mouse fibroblasts)	Visible light irradiation (Hg lamp)	Low toxicity in the dark and higher toxicity in the presence of light	[42]
Sf	Bovine Serum Albumin pH 7.4 (in Tris–HCl-NaOH buffer solution)	Study of the damage caused by ROS in bovine serum albumin	Controllable Serial-Ultrasonic apparatus (frequency 59 kHz and power 50 W	Damage of Bovine Serum Albumin under ultrasonic irradiation in the presence of Sf.	[96]
Sf	Aqueous solution (PBS containing 7% ethanol)	Oral-pathogenic species (Gram-positive and Gram-negative)	Laser Light (20 J cm^{-2})	Significant antibacterial impact on different oral pathogenic species	[97]

Compound	Delivery system	Application	Illumination condition	Effect	Reference
Sf	Gold nanoparticles incorporated into a copolymer emulsion	*Staphylococcus aureus* and *Escherichia coli*,	28 W white light source	The bacterial count reduced (>4 ± 0.3 log kill)	[43]
NR Monob rominated NR	Aqueous solution (in PBS pH 7.4)	*Staphylococcus aureus*	Light dose 7.6–30.2 J cm^{-2}	Photoantimicrobial effect (>3 log_{10}) killing	[98]
Sf	Aqueous solution	*Hyalomma dromedarii*	Spot white-light source (power 100 W)	Reduction of ovipositing, eggs per female, tickets laying viable eggs and hatched eggs	[99]
NR	Gold Nanoparticles and Sodium thioglycolate	NIH-3 T3 fibroblast (noncancerous) and 4 T1 tumor cell lines	Twin Flex Laser LED MM optics λ = 440 nm; 220 mW	Reduction of cell viability	[22]
PhS-Chlorambucil conjugate	Encapsulated liposomes (DMPC)	HeLa (human, cervical carcinoma)		Excellent cell contrast facilitating its use as a theranostic anti-cancer drug	[23]
Sf	Pluronic® F127 and P123 4% (w/V)	*In vitro*: *Staphylococcus aureus*, *Escherichia coli*, *Streptococcus agalactiae*, *Corynebacterium bovis* *In vivo*: prevent mastitis in Dutch dairy cows.	*In vitro* studies: Green LED λ = 520 nm (7.2 mW cm^{-2}) *In vivo* studies: Green LED λ = 520 nm (12.7 mW cm^{-2})	*In vitro* studies of the Sf-F127 and Sf-P123 systems proved to be efficient in inactivating the bacteria that cause bovine mastitis. *In vivo* studies prevent bovine mastitis	[33]
NR Mono brominated NR	Aqueous solution (in PBS pH 7.4)	*Staphylococcus aureus*	Parathom lamp (OSRAM-5 W) 8.4 mW cm^{-2} (15–30 min)	2–3 log of killing	[98]
Sf	Stimuli-responsive hydrogel by F127 Pluronic® and Carbopol (C934P)	Mastitis treatment	*In vitro* studies: Green LED λ = 520 nm (7.2 mW cm^{-2}) *In vivo* studies: Green LED λ = 520 (12.7 mW cm^{-2})	*In vitro*: efficiency in the inactivation of pathogens that cause mastitis. *In vivo*: Sf is highly efficient for mastitis treatment.	[100]
Sf	Pluronic® F127 4% (w/V)	*In vitro*: *Staphylococcus aureus*, *Escherichia coli*	Green LED λ = 520 (7.2 mWcm^{-2})	The bacteria were sensitive to the photodynamic action of Sf	[101]
PhS polyhedral oligomeric silsesquioxane	Aqueous solution (in PBS pH 7.4)	*Staphylococcus aureus* and *Escherichia coli*	LED-based light source λ = 522 nm (10.6 mW cm^{-2})	Bacterial cells were eradicated by ROS produced upon irradiation	[102]

Table 2.
Studies of phenazine class compounds in photodynamic applications.

Table 2 shows a few amounts of work on this class of compounds and the way performed majority *in vitro*, which the scientific community can still better explore for applications in PDT. Combining the biocompatibility of the copolymer micelles and the non-toxicity of phenazines compounds, the continuity of pre-clinical studies for developing new photoactive-based phenazine formulations has been motivated for applications in different species of animals. These formulations can treat a wide range of diseases associated with different types of pathogenic microorganisms such as bacteria, fungi, viruses, and parasites. In this way, recent research has still been developed for the first time *in vivo*. It is in the submission phase of (unpublished) promising results concerning the prevention and treatment of mastitis. This review seeks to address the lack of literature on the approach of phenazines as a potential PS for PDT application. It is hoped that this work could offer some valuable information in developing new types of DDS systems.

7. Conclusion

Several products are used to control microorganisms and treat diseases like cancer; however, they present disadvantages such as cost and lack of effectiveness against resistant microorganisms. Photosensitization has recently gained attention benefiting from the use of a wide range of PS associated with a light source. The oxygen species produced by photoexcitation of a PS attack cancer cells or microorganisms non-selectively. Phenazine compounds showed promising phototoxicity, and their applicability has still been poorly investigated in the prevention and treatment of diseases caused by microorganisms and diseases related to abnormal cell growth, such as cancer. Furthermore, incorporating PS into polymeric micelles produced efficient, biocompatible formulations with better stability than aqueous systems.

Conflict of interest

The authors declare no conflict of interest.

Appendices and nomenclature

PDT	Photodynamic Therapy
aPDT	antimicrobial Photodynamic Therapy
PS	photosensitizing compounds
ROS	reactive oxygen species
$^3O_2$	molecular oxygen
$^1O_2$	singlet oxygen
O_2^-	superoxide
H_2O_2	hydrogen peroxide
$\Phi_\Delta {}^1O_2$	singlet oxygen quantum yield
LDL	low-density proteins
FDA	Food and Drug Administration
DDS	drug delivery systems
IC	internal conversion
ISC	intersystem crossing

CMC	critical micellar concentration
CMT	critical micellar temperature
HOMO	Highest Occupied Molecular Orbital
LUMO	Lowest Unoccupied Molecular Orbital

Author details

Ranulfo Combuca da Silva Junior[1*], Katieli da Silva Souza Campanholi[1], Flávia Amanda Pedroso de Morais[2], Laura Adriane de Moraes Pinto[3], Fabiana dos Santos Rando[4], Magali Soares dos Santos Pozza[5] and Wilker Caetano[1]

1 Department of Chemistry, State University of Maringá, Maringá, Paraná, Brazil

2 Department of Basic Health Sciences, State University of Maringá, Maringá, Paraná, Brazil

3 Animal Science Post-Graduate Program, Federal University of Paraná- Sector Palotina, Palotina, Paraná, Brazil

4 Agronomy Post-Graduate Program, State University of Mato Grosso do Sul, Cassilândia, Mato Grosso do Sul, Brazil

5 Department of Animal Science, State University of Maringá, Maringá, Paraná, Brazil

*Address all correspondence to: rcsjunior@uem.br

References

[1] Simplicio FI, Maionchi F, Hioka N. Terapia fotodinâmica: aspectos farmacológicos, aplicações e avanços recentes no desenvolvimento de medicamentos. Química Nova. 2002;**25**:801-807. DOI: 10.1590/S0100-40422002000500016

[2] Dougherty TJ, Gomer CJ, Henderson BW, Jori G, Kessel D, Korbelik M, et al. Photodynamic therapy. JNCI: Journal of the national cancer institute. 1998;**90**:889-905. DOI: 10.1093/JNCI/90.12.889

[3] Bonin E, dos Santos AR, da Silva AF, Ribeiro LH, Favero ME, Campanerut-Sá PAZ, et al. Photodynamic inactivation of foodborne bacteria by eosin Y. Journal of Applied Microbiology. 2018;**124**:1617-1628. DOI: 10.1111/jam.13727

[4] de Morais FAP, Gonçalves RS, Braga G, Calori IR, Pereira PCS, Batistela VR, et al. Stable dipalmitoylphosphatidylcholine liposomes coated with an F127 copolymer for hypericin loading and delivery. ACS Applied Nano Materials. 2020;**3**:4530-4541. DOI: 10.1021/acsanm.0c00386

[5] de Moeais FAP, Gonçalves RS, Campanholi KS, França BM, Capeloto OA, Lazarin-Bidoia D, et al. Photophysical characterization of hypericin-loaded in micellar, liposomal and copolymer-lipid nanostructures based F127 and DPPC liposomes. Spectrochimica Acta Part A: Molecular and Biomolecular Spectroscop. 2021;**248**:119173-112021. DOI: 10.1016/J.saa.2020.1419173

[6] de Souza PCP, Costa PFA, Pellosi DS, Clori IR, Vilsinski BH, Estevão BM, et al. Photophysical properties and interaction studies of rose Bengal derivatives with biomimetic systems based in micellar aqueous solutions. Journal of Molecular Liquids. 2017;**230**:674-685. DOI: 10.1016/j.molliq.2017.01.055

[7] Machado AEDH. Terapia fotodinâmica: princípios, potencial de aplicação e perspectivas. Química Nova. 2000;**23**:237-243. DOI: 10.1590/S0100-40422000000200015

[8] Deeb D, Xu YX, Jiang H, Gao X, Janakiraman N, Chapman RA, et al. Curcumin (diferuloyl-methane) enhances tumor necrosis factor-related apoptosis-inducing ligand-induced apoptosis in lncap prostate cancer cells[1]. Molecular Cancer Therapeutics. 2003;**2**:95-103

[9] Zhu TC, Finaly JC. Prostate PDT dosimetry. Photodiagnosis and Photodynamic Therapy. 2006;**3**:234-246. DOI: 10.1016/j.pdpdt.2006.08.002

[10] Bastos MM, Boechat N, Gomes ATPC, Neves AG, Cavaleiro J. O uso de porfirinas em terapia fotodinâmica no tratamento da leishmaniose cutânea. Revista Virtual de Química. 2012;**4**:257-267. DOI: 10.5935/1984-6835.20120021

[11] de Oliveira EL, Ferreira SBS, Castro-Hoshino LV, Campanholi KSS, Calori IR, de Morais FAP, et al. Thermoresponsive hydrogel-loading Aluminum chloride Phthalocyanine as a drug release platform for topical administration in photodynamic therapy. Langmuir. 2021;**37**:3202-3213. DOI: 10.1021/acs.langmuir.1c00148

[12] Woodcock J, Griffin JP, Behrman RE. Development of novel combination therapies. The New England Journal of Medicine. 2011;**364**:985-987. DOI: 10.1056/NEjMp1101548

[13] de Oliveira KT, Souza JM, Globo NRS, Assis FF, Brocksom TJ. Conceitos fundamentais e Aplicações de Fotossensibilizadores do tipo Porfirinas, clorinas e Ftalocianinas em Terapias Fotônicas. Revista Virtual de Química. 2015;7:310-335. DOI: 10.5935/1984-6835.20150016

[14] Ormond AB, Freeman HS. Dye sensitizers for photodynamic therapy. Materials. 2013;**6**:817-840. DOI: 10.3390/ma6030817

[15] Yoon I, Li JZ, Shim YK. Advance in photosensitizers and light delivery for photodynamic therapy. Clinical Endoscopy. 2013;**46**:7-23. DOI: 10.5946/ce.2013.46.1.7

[16] Braga G, Campanholi KSS, Ferreira SBS, Calori IR, Oliveira JH, Bruschi ML, et al. Tautomeric and aggregational dynamics of curcumin-supersaturated pluronic nanocarriers. ACS Applied Polymer Materials. 2020;**2**:4493-4511. DOI: 10.1021/acsapm.0c00589

[17] Melo WC, Perussi JR. Comparando inativação fotodinâmica e antimicrobianos. Revista de Ciências Farmacêuticas Básica e Aplicada. 2012;**33**:331-340

[18] Campanholi KSS, Zanqui AB, de Morais FAP, Jaski JM, Gonçalves RS, Silva-Junior RC, et al. Obtaining phytotherapeutic chlorophyll extracts using pressurized liquid technology. The Journal of Supercritical Fluids. 2022;**180**:105457. DOI: 10.1016/j.supflu.2021.105457

[19] Byvaltsev VA, Bardonova LA, Onaka NR, Polkin RA, Ochkal SV, Shepelev VV, et al. Acridine orange: A review of novel applications for surgical cancer imaging and therapy. Frontiers in Oncology. 2019;**9**:925. DOI: 10.3389/fonc.2019.00925

[20] Carobeli LR, Meirelles LEF, Damke GMZF, Damke E, Souza MVF, Mari NL, et al. Phthalocyanine and its formulations: A promising photosensitizer for cervical Cancer phototherapy. Pharmaceutics. 2021;**13**:2057. DOI: 10.3390/pharmaceutics13122057

[21] Wiench R, Skaba D, Matys J, Grzech-Lesniak K. Efficacy of toluidine blue-mediated antimicrobial photodynamic therapy on *Candida spp.* A systematic review. Antibiotics. 2021;**10**:349. DOI: 10.3390/antibiotics10040349

[22] Verissimo TV, Santos NT, Silva JR, Azevedo RB, Gomes AJ, Lunardi CN. In vitro cytotoxicity and phototoxicity of surface-modified gold nanoparticles associated with neutral red as a potential drug delivery system in phototherapy. Materials Science and Engineering: C. 2016;**65**:199-204. DOI: 10.1016/j.msec.2016.04.030

[23] Miksa B, Sierant M, Slorupska E, Michalski A, Kazmierski S, Steinke U, et al. Chlorambucil labelled with the phenosafranin scaffold as a new chemotherapeutic for imaging and cancer treatment. Colloids and Surfaces. B, Biointerfaces. 2017;**159**:820-828. DOI: 10.1016/J.COLSURFB.2017.08.045

[24] Jankowski A, Jankowski S, Mironczyk A, Niedbach J. The action of photosensitizers and serum in a bactericidal process. II. The effects of dyes: Hypericin, eosin Y and saphranine O. Polish Journal of Microbiology. 2005;**54**:323-330

[25] Vilsinsk BH, Witt MA, Barbosa PM, Montanha MC, Nunes CS, Bellettini IC, et al. Formulation of chloroaluminum phthalocyanine incorporated into PS-b-PAA diblock copolymer nanomicelles. Journal of Molecular Liquids. 2018;**271**:949-958. DOI: 10.1016/j.molliq.2018.09.034

[26] Rossin ARS, de Oliveira EL, de Morais FAP, Silva-Junior RC, Scheid DT, Caetano W, et al. Terapia Fotodinâmica em Eletrofiação: Revisão de técnicas e aplicações. Química Nova. 2020;**43**:613-622. DOI: 10.21577/0100-4042.20170524

[27] Lakowicz JR. Principles of Fluorescence Spectroscopy. Boston, US: Springer Science + Business Media, LLC; 2006. ISBN: 978-0-387-46312-4

[28] Agostinis P, Berg K, Cengel KA, Foster TH, Girotti AW, Gollnick SO, et al. Photodynamic therapy of cancer: An update. CA: A Cancer Journal for Clinicians. 2011;**61**:250-281. DOI: 10.3322/CAAC.20114

[29] Dai T, Huang Y-Y, Hamblin MR. Photodynamic therapy for localized infections state of the art. Photodiagnosis and Photodynamic Therapy. 2009;**6**:170-188. DOI: 10.1016/j.pdpdt.2009.10.008

[30] Kubler AC. Photodynamic therapy. Medical Laser Application. 2005;**20**:37-45. DOI: 10.1016/j.mla.2005.02.001

[31] Kharkwal GB, Sharma SK, Huang Y-Y, Dai T, Hamblin MR. Photodynamic therapy for infections: Clinical applications. Lasers in Surgery and Medicine. 2011;**43**:755-767. DOI: 10.1002/LSM.21080

[32] Saraiva BB, Rodrigues BM, Silva-Junior RC, Scapim MRS, Lancheros CAC, Nakamura CV, et al. Photodynamic inactivation of *Pseudomonas fluorescens* in Minas Frescal cheese using curcumin as a photosensitizer. LWT. 2021;**151**:112143. DOI: 10.1016/j.lwt.2021.112143

[33] da Silva-Junior RC, Campanholi KSS, de Morais FAP, Pozza MSS, Santos GT, Hioka N, et al. Development and applications of safranine-loaded Pluronic® F127 and P123 photoactive nanocarriers for prevention of bovine mastitis: *In vitro* and *in vivo* studies. Dyes and Pigments. 2019;**167**:204-215. DOI: 10.1016/j.dyepig.2019.04.037

[34] Kataoka K, Harada A, Nagaski Y. Block copolymer micelles for drug delivery: Design, characterization and biological significance. Advanced Drug Delivery Reviews. 2012;**64**:37-48. DOI: 10.1016/j.addr.2012.09.013

[35] Rijcken CJF, Hofman JW, van Zeeland F, Hennink WE, van Nostrum CF. Photosensitiser-loaded biodegradable polymeric micelles: Preparation, characterisation and *in vitro* PDT efficacy. Journal of Controlled Release. 2007;**124**:144-153. DOI: 10.1016/j.conrel.2007.09.002

[36] de Morais FAP, Emuno-Junior A, Gonçalves RS, Cesar GB, Miranda N, Vilsinski BH, et al. Hypericin photodynamic activity. Part III: in vitro evaluation in different nanocarriers against trypomastigotes of *Trypanosoma cruzi*. Photochemical & Photobiological Sciences. 2019;**18**:487-494. DOI: 10.1039/c8pp00444g

[37] Li Y, Jang WD, Nishiyama N, Kishimura A, Kawauchi S, Morimoto Y, et al. Dendrimer generation effects on photodynamic efficacy of dendrimer porphyrins and dendrimer-loaded supramolecular nanocarriers. Chemistry of Materials. 2007;**19**:5557-5562. DOI: 10.1021/cm071451m

[38] Mihoub AB, Larue L, Moussaron A, Youssef Z, Colombeau L, Baros F, et al. Use of cyclodextrins in anticancer photodynamic therapy treatment. Molecules. 2018;**23**:1936. DOI: 10.339/molecules23081936

[39] Semeraro P, Chimient G, Altamura E, Fini P, Rizzi V, Cosma P. Chlorophyll a in cyclodextrin supramolecular complexes as a natural

photosensitizer for photodynamic therapy (PDT) applications. Materials Science and Engineering: C. 2018;**85**:47-56. DOI: 10.1016/j.msec.2017.12.012

[40] Wang J, Liu L, Chen J, Deng M, Feng X, Chen L. Supramolecular nanoplatforms via cyclodextrin host-guest recognition for synergistic gene-photodynamic therapy. European Polymer Journal. 2019;**118**:222-230. DOI: 10.1016/jeurpolymj.2019.04.051

[41] Campanholi KSS, Silva-Junior RC, Silva IC, Santos RS, Vecchi CF, Bruschi ML, et al. Stimulus-responsive phototherapeutic micellar platform of rose Bengal B: A new perspective for the treatment of wounds. Journal of Drug Delivery Science and Technology. 2021;**2021**:102739. DOI: 10.1016/j.jddst.2021.102739

[42] do Nascimento GM, Oliveira RC, Pradie NA, PRG L, Worfel PR, Martinez GR, et al. Single-wall carbon nanotubes modified with organic dyes: Synthesis, characterization and potential cytotoxic effects. Journal of Photochemistry and Photobiology A: Chemistry. 2010;**211**:99-107. DOI: 10.1016/j.jphotochem.2010.01.019

[43] Beech SJ, Noimark S, Page K, Noor N, Allan E, Parkin IP. Incorporation of crystal violet, methylene blue and safranin O into a copolymer emulsion; the development of a novel antimicrobial paint. RSC Advances. 2015;5:26364-26375. DOI: 10.1039/c5ra01673h

[44] Drogat N, Granet R, Morvan CL, Begaud-Grimaud G, Krausz P, Sol V. Chlorin-PEI-labeled cellulose nanocrystals: Synthesis, characterization and potential application in PDT. Bioorganic & Medicinal Chemistry Letters. 2012;**22**:3648-3652. DOI: 10.1016/j.bmcl.2012.04.044

[45] Jiang S, Zhu R, He X, Wang J, Wang M, Qian Y, et al. Enhanced photocytotoxicity of curcumin delivered by solid lipid nanoparticles. International Journal of Nanomedicine. 2017;**12**:167-178. DOI: 10.2147/ijn.s123107

[46] Bechet D, Couleaud P, Frochot C, Virot ML, Guillemin F, Barberi-Heyob M. Nanoparticles as vehicles for delivery of photodynamic therapy agents. Trends in Biotechnology. 2008;**26**:612-621. DOI: 10.1016/j.tibtech.2008.07.007

[47] Malmsten M. Surfactants and Polymers in Drug Delivery. Boca Raton, US: CRC Press; 2002. DOI: 10.1201/9780824743758

[48] van Nostrum CF. Polymeric micelles to deliver photosensitizers for photodynamic therapy. Advanced Drug Delivery Reviews. 2004;**56**:9-16. DOI: 10.1016/J.ADDR.2003.07.013

[49] Batrakova EV, Kabavov AV. Pluronic block copolymers: Evolution of drug delivery concept from inert nanocarriers to biological response modifiers. Journal of Controlled Release. 2008;**130**:98-106. DOI: 10.1016/j.jconrel.2008.04.013

[50] Senapati S, Mahanta AK, Kumar S, Maiti P. Controlled drug delivery vehicles for cancer treatment and their performance. Signal Transduction and Targeted Therapy. 2018;**3**:1-19. DOI: 10.1038/s41392-017-0004-3

[51] Lima AM, Pizzol CD, Monteiro FBF, Creczynski-Pasa TB, Andrade GP, Ribeiro AO, et al. Hypericin encapsulated in solid lipid nanoparticles: Phototoxicity and photodynamic efficiency. Journal of Photochemistry and Photobiology B: Biology. 2013;**125**:146-154. DOI: 10.1016/j.jphotobiol.2013.05.010

[52] Campanholi KSS, Braga G, Rocha NL, Francisco LMB, Oliveira EL, Bruschi ML, et al. Biomedical platform development of a chlorophyll-based

extract for topic photodynamic therapy: Mechanical and spectroscopic properties. Langmuir. 2018;**34**:8230-8244. DOI: 10.1021/acs.langmuir.8b00658

[53] Gerola A, Costa PF, de Morais FA, Tsubone TA, Calared AO, Nakamura CV, et al. Liposome and polymeric micelle-based delivery systems for chlorophylls: Photodamage effects on *Staphylococcus aureus*. Colloids and Surfaces B: Biointerfaces. 2019;**177**:487-495. DOI: 10.1016/j.colsurfb.2019.02.032

[54] George A. Advances in biomimetics. BoD–Books on Demand. Rijeka, Croatia: InTech; 2011. ISBN 978-953-307-191-6

[55] Neuman MG, Gessner F, Oliveira VA. Formation of mixed dimers in solutions of basic dyes. Journal of the Chemical Society, Faraday Transactions. 1990;**86**:3551-3555. DOI: 10.1039/ft9908603551

[56] Pastre IA, Neumann MG. Protonation of the triplet state of safranine in the presence of cationic micelles and polyelectrolytes. Journal of Photochemistry and Photobiology, A: Chemistry. 1994;**79**:115-119. DOI: 10.1016/1010-6030(93)03744-2

[57] Bozkurt E, Bayraktutan T, Mahmut A, Toprak M. Spectroscopic studies on the interaction of fluorescein and safranine T in PC liposomes. Spectrochimica Acta Part A: Molecular and Biomolecular Spectroscopy. 2013;**101**:31-35. DOI: 10.1016/j.saa.2012.09.082

[58] Junqueira MV, Borghi-Pangoni FB, Ferreira SBS, Rabello BR, Hioka N, Bruschi ML. Functional polymeric systems as delivery vehicles for methylene blue in photodynamic therapy. Langmuir. 2016;**32**:19-27. DOI: 10.1021/acs.langmuir.5b02039

[59] Ferreira SBDS, Silva JB, Junqueira MV, Borghi-Pangonia Raquel FB, Gomes RG, Bruschi ML, et al. The importance of the relationship between mechanical analyses and rheometry of mucoadhesive thermoresponsive polymeric materials for biomedical applications. Journal of the Mechanical Behavior of Biomedical Materials. 2017;**74**:142-153. DOI: 10.1016/j.jmbbm.2017.05.040

[60] Trigo-Gutierrez JK, Zanatta GC, Ortega ALM, Balastegui MIC, Sanita PV, Pavarina AC, et al. Encapsulation of curcumin in polymeric nanoparticles for antimicrobial Photodynamic Therapy. PLoS One. 2017;**12**:e0187418. DOI: 10.1371;journal.pone.0187418

[61] Gerola AP, de Morais FAP, Costa PFA, Kimura E, Caetano W, Hioka N. Characterization of chlorophyll derivatives in micelles of polymeric surfactants aiming photodynamic applications. Spectrochimica Acta Part A: Molecular and Biomolecular Spectroscopy. 2017;**173**:213-221. DOI: 10.1016/j.saa.2016.09.019

[62] Vukicevic M, Hegge AB, Vulic P, Tonnesen H. Poloxamer-based curcumin solid dispersions for ex tempore preparation of supersaturated solutions intended for antimicrobial photodynamic therapy. Pharmaceutical Development and Technology. 2015;**20**:863-871. DOI: 10.3109/10837450.2014.930489

[63] Costa PFA, Gerola AP, Pereira PCS, Souza BS, Caetano W, Fiedler HD, et al. Chlorophylls B formulated in nanostructured colloidal solutions: Interaction, spectroscopic, and photophysical studies. Journal of Molecular Liquids. 2019;**274**:393-401. DOI: 10.1016/j.molliq.2018.10.143

[64] Alexandridis P, Holzwarth JF, Hatton TA. Micellization of poly (ethylene oxide)-poly (propylene oxide)-poly (ethylene oxide) triblock

copolymers in aqueous solutions: Thermodynamics of copolymer association. Macromolecules. 1994;**27**:2414-2425

[65] Freitas CF, Rocha NL, Pereverzieff IS, Batistela VR, Malacarne LC, Hioka N, et al. Potential of triblock copolymers Pluronic® P-84 and F-108 with erythrosine B and its synthetic ester derivatives for photodynamic applications. Journal of Molecular Liquids. 2021;**322**:114904. DOI: 10.1016/j.molliq.2020.114904

[66] Alizadeh AM, Sadeghizadeh M, Najafi F, Ardestani SK, Erfani-Moghadam V, Khanik M, et al. Encapsulation of curcumin in diblock copolymer micelles for cancer therapy. BioMed Research International. 2015;**2015**:1-13. DOI: 10.1155/2015/824746

[67] Obata M, Masuda S, Takahashi M, Yazaki K, Hirohara S. Effect of the hydrophobic segment of an amphiphilic block copolymer on micelle formation, zinc phthalocyanine loading, and photodynamic activity. European Polymer Journal. 2021;**147**:10325. DOI: 10.1016/j.eurpolymj.2021.110325

[68] Mavrodi DV, Bonsall RF, Delaney SM, Soule MJ, Philips G, Thomashow LS. Functional analysis of genes for biosynthesis of pyocyanin and phenazine-1-carboxamide from *Pseudomonas aeruginosa* PAO1. Journal of Bacteriology. 2001;**183**:6454-6465. DOI: 10.1128/jb.183.21.6454-6465.2001

[69] Yan J, Liu W, Cai J, Wang Y, Li D, Hua H, et al. Advances in Pphenazines over the past decade: Review of their pharmacological activities, mechanisms of action, biosynthetic pathways and synthetic strategies. Marine Drugs. 2021;**19**:610. DOI: 10.3390/md19110610

[70] Stover CK, Pham XQ, Erwin AL, Mizoguchi SD, Warrener P, Hickey MJ, et al. Complete genome sequence of *Pseudomonas aeruginosa* PAO1, an opportunistic pathogen. Nature. 2000;**406**:959-964

[71] Abu EA, Su S, Sallans L, Boissy RE, Greatens A, Heineman WR, et al. Cyclic voltammetric, fluorescence and biological analysis of purified aeruginosin a, a secreted red pigment of *Pseudomonas aeruginosa* PAO1. Microbiology. 2013;**159**:1736-1747. DOI: 10.1099/mic.0.065235-0

[72] Kennedy RK, Veena V, Naik PR, Lakshmi P, Krishna R, Sudharani S, et al. Phenazine-1-carboxamide (PCN) from *Pseudomonas sp.* strain PUP6 selectively induced apoptosis in lung (A549) and breast (MDA MB-231) cancer cells by inhibition of antiapoptotic Bcl-2 family proteins. Apoptosis. 2015;**20**:858-868. DOI: 1007/s10495-015-1118-0

[73] Mahan AM, Ragab SS, Hashem A, Ali MM, Nada AA. Synthesis and antiproliferative activity of novel polynuclear heterocyclic compounds derived from 2, 3-diaminophenazine. European Journal of Medicinal Chemistry. 2015;**90**:568-576. DOI: 10.1016/j.ejmech.2013.12.007

[74] Moris MA, Andrieu C, Rocchi P, Seillan C, Acunzo J, Brunel F, et al. 2, 3-Dialkoxyphenazines as anticancer agents. Tetrahedron Letters. 2015;**56**:2695-2698. DOI: 10.1016/j.tetlet.2015.04.003

[75] Reddy MV, Valasani KR, Lim KT, Jeong YT. Tetrame thylguanidiniumchlorosulfonate ionic liquid (TMG IL): An efficient reusable catalyst for the synthesis of tetrahydro-1 H-benzo [a] chromeno [2, 3-c] phenazin-1-ones under solvent-free conditions and evaluation for their *in vitro* bioassay

activity. New Journal of Chemistry. 2015;**39**:9931-9941. DOI: 10.1039/C5NJ01866H

[76] Lu Y, Wang L, Wang X, Xi T, Liao J, Wang Z, et al. Design, combinatorial synthesis and biological evaluations of novel 3-amino-1′-((1-aryl-1H-1, 2, 3-triazol-5-yl) methyl)-2′-oxospiro [benzo [a] pyrano [2, 3-c] phenazine-1, 3′-indoline]-2-carbonitrile antitumor hybrid molecules. European Journal of Medicinal Chemistry. 2017;**135**:125-141. DOI: 10.1016/j.ejmech.2017.04.040

[77] Cimmino A, Evidente A, Mathieu V, Andolfi A, Lefranc F, Kornienco A, et al. Phenazines and cancer. Natural Product Reports. 2012;**29**:487-501. DOI: 10.1039/c2np00079b

[78] Pachon OG, Azqueta A, Lavaggi ML, Cerani AL, Creppy E, Collins A, et al. Antitumoral effect of phenazine N 5, N 10-dioxide derivatives on Caco-2 cells. Chemical Research in Toxicology. 2008;**21**:1578-1585. DOI: 10.1021/tx80002k

[79] Lavaggi ML, Aguirre G, Boiani L, Orelli L, Garcia B, Cerecetto H, et al. Pyrimido [1, 2-a] quinoxaline 6-oxide and phenazine 5, 10-dioxide derivatives and related compounds as growth inhibitors of Trypanosoma cruzi. European Journal of Medicinal Chemistry. 2008;**43**:1737-1741. DOI: 10.1016/j.ejmech.2007.10.031

[80] Cardozo VF, Oliveira AG, Nishio EK, Perugini MRE, Andrade CGTJ, Silveira WD, et al. Antibacterial activity of extracellular compounds produced by a *Pseudomonas* strain against methicillin-resistant *Staphylococcus aureus* (MRSA) strains. Annals of Clinical Microbiology and Antimicrobials. 2013;**12**:1-8

[81] Gorantla JN, Kumar SN, Nisha GV, Sumandu AS, Dileep C, Sudaresan A, et al. Purification and characterization of antifungal phenazines from a fluorescent *Pseudomonas* strain FPO4 against medically important fungi. Journal de mycologie médicale. 2014;**24**:185-192. DOI: 10.1016/j.mycmed. 2014.02.003

[82] Tupe SG, Kulkarni RR, Shirazi F, Sant DG, Joshi SP, Deshpande MV. Possible mechanism of antifungal phenazine-1-carboxamide from *Pseudomonas sp.* against dimorphic fungi B enjaminiella poitrasii and human pathogen *Candidda albicans*. Journal of Applied Microbiology. 2015;**118**:39-48. DOI: 10.1111/jam.12675

[83] Cha JW, Lee S, Kim MC, Thida M, Lee JW, Park JS, et al. Pontemazines a and B, phenazine derivatives containing a methylamine linkage from *Streptomyces sp.* UT1123 and their protective effect to HT-22 neuronal cells. Bioorganic & Medicinal Chemistry Letters. 2015;**25**:5083-5086. DOI: 10.1016/j.bmcl. 2015.10.019

[84] Yao BL, Mai YW, Chen SB, Xie HT, Yao PF, Ou TM, et al. Design, synthesis and biological evaluation of novel 7-alkylamino substituted benzo [a] phenazin derivatives as dual topoisomerase I/II inhibitors. European Journal of Medicinal Chemistry. 2015;**92**:540-553. DOI: 10.1016/j. ejmech.2015.01.024

[85] Kennedy RK, Naik PR, Venna V, Lakshmi BS, Lakshmi P, Krishna R, et al. 5-methyl phenazine-1-carboxylic acid: A novel bioactive metabolite by a rhizosphere soil bacterium that exhibits potent antimicrobial and anticancer activities. Chemico-Biological Interactions. 2015;**231**:71-82. DOI: 10.1016/j.cbi. 2015.03.002

[86] Varsha KK, Nishant G, Sneha SM, Shilpa G, Devendra L, Priya S, et al. Antifungal, anticancer and

aminopeptidase inhibitory potential of a phenazine compound produced by *Lactococcus* BSN307. Indian Journal of Microbiology. 2016;**56**:411-416. DOI: 10.1007/s12088-016-0597-1

[87] Ali HM, Shikh MS, Salem MZ. Isolation of bioactive phenazine-1-carboxamide from the soil bacterium Pantoea agglomerans and study of its anticancer potency on different cancer cell lines. Journal of AOAC International. 2016;**99**:1233-1239. DOI: 10.5740/jaoacint.16-0090

[88] Makgatho E, Mbajiorgu E. In vitro investigation of clofazimine analogues for antiplasmodial, cytotoxic and pro-oxidative activities. African Health Sciences. 2017;**17**:191-198. DOI: 10.4314/ahs.v17i1.24

[89] Krishnaiaha M, Almeida NR, Udumula V, Song Z, Chhonker YS, Abdelmoaty MM, et al. Synthesis, biological evaluation, and metabolic stability of phenazine derivatives as antibacterial agents. European Journal of Medicinal Chemistry. 2018;**143**:936-947. DOI: 10.1016/j.ejmech.2017.11.026

[90] Liao J, Wang L, Wu Z, Wan Z, Chen J, Zhong Y, et al. Identification of phenazine analogue as a novel scaffold for thioredoxin reductase I inhibitors against Hep G2 cancer cell lines. Journal of Enzyme Inhibition and Medicinal Chemistry. 2019;**34**:1158-1163. DOI: 10.1080/14756366.2019.1624541

[91] Gu P-Y, Zhao Y, He JH, Zhang J, Wang C, Xu Q-F, et al. Synthesis, physical properties, and light-emitting diode performance of phenazine-based derivatives with three, five, and nine fused six-membered rings. The Journal of organic chemistry. 2015;**80**:3030-3035. DOI: 10.1021/jo5027707

[92] Liu X, Yang Z, Xu W, Chu Y, Yang J, Yan Y, et al. Fine tuning of pyridinium-functionalized dibenzo [a, c] phenazine near-infrared AIE fluorescent biosensors for the detection of lipopolysaccharide, bacterial imaging and photodynamic antibacterial therapy. Journal of Materials Chemistry C. 2019;7:12509-12517. DOI: 10.1039/C9TC04427B

[93] Udumula V, Endres JL, Harper CN, Jaramillo L, Zhong HA, Bayles KW, et al. Simple synthesis of endophenazine G and other phenazines and their evaluation as anti-methicillin-resistant *Staphylococcus aureus* agents. European Journal of Medicinal Chemistry. 2017;**125**:710-721. DOI: 10.1016/j.ejmech.2016.09.079

[94] Elhadyv HA, Mekawy RE, Fadda A. Valuable chemistry of phenazine derivatives: Synthesis, reactions and, applications. European Journal of Medicinal Chemistry. 2020;**12**:2894-2926. DOI: 10.1080/10406638.2020.1833051

[95] Sasnauskiene A, Kadziauskas J, Vezelyte N, Jonusiene V, Kirveliene V. Apoptosis, autophagy and cell cycle arrest following photodamage to mitochondrial interior. Apoptosis. 2009;**14**:276-286. DOI: 10.1007/s10495-008-0292-8

[96] Li J, Wang Z, Wang J, Gao J, Zou M, Li Y, et al. Spectroscopic investigation on the sonodynamic activity of Safranine T to bovine serum albumin damage. Journal of Luminescence. 2012;**132**:282-288. DOI: 10.1016/jlumin.2011.09.014

[97] Voos AC, Krans DMD, Tonndorf-martini S, Dipl-Ing AV, Sigusch H, Staudte H, et al. Photodynamic antimicrobial effect of safranine O on an *ex vivo* periodontal biofilm. Lasers in Surgery and Medicine. 2014;**46**:235-243. DOI: 10.1002/Ism.22217

[98] Urrutia MN, Ortiz CS, Alovero FL. Mechanistic insight into the

photodynamic effect mediated by neutral red and a new Azine compound in *Staphylococcus aureus* cells. Chemistry & Biodiversity. 2019;**16**:e1900262. DOI: 10.1002/cbdv.201900262

[99] Khater H, Hendawy N, Govindarajan M, Marugan K, Banelli G. Photosensitizers in the fight against ticks: Safranin as a novel photodynamic fluorescent acaricide to control the camel tick *Hyalomma dromedarii* (Ixodidae). Parasitology Research. 2016;**15**:3747-3758. DOI: 10.1007/s00436-016-5136-9

[100] da Silva-Junior RC, Campanholi KSS, de Morais FAP, Pozza MSS, Castro-Hoshino LV, Baesso ML, et al. Photothermal stimuli-responsive hydrogel containing Safranine for mastitis treatment in veterinary using phototherapy. ACS Applied Bio Materials. 2020;**4**:581-596. DOI: 10.1021/acsabm.0.c01143

[101] da Silva-Junior RC, Campanholi KSS, Rando FS, Gonçalves RS, Lazarin-Bidóia D, de Morais FAP, et al. Antimicrobial photoinactivation approach based on Safranine-O loaded F127 copolymeric micelles for control of gram-negative and gram-positive bacteria. Dyes and Pigments. 2021;**197**:109900. DOI: 10.1016/j.dyepog.2021.109900

[102] Rozga-Wijas K, Bak-Sypien I, Turecka K, Narajczyk M, Waleron K. Cationic Phenosafranin photosensitizers based on polyhedral Oligomeric Silsesquioxanes for inactivation of gram-positive and gram-negative Bacteria. International Journal of Molecular Sciences. 2021;**22**:13373. DOI: 10.3390/ijms222413373

[103] Urrutia MN, Alovero FL, Ortiz CS. New azine compounds as photoantimicrobial agents against *Staphylococcus aureus*. Dyes and Pigments. 2015;**116**:27-35. DOI: 10.1016/j.dyepig.2014.12.021

[104] Gopidas K, Kamat PV. Photophysics and photochemistry of phenosafranin dye in aqueous and acetonitrile solutions. Journal of Photochemistry and Photobiology, A: Chemistry. 1989;**48**:291-301. DOI: 10.1016/1010-6030(89)87010-8

[105] Jockusch S, Tempo H-J, Schnabel W, Turro NJ. Photoinduced energy and electron transfer between ketone triplets and organic dyes. The Journal of Physical Chemistry A. 1997;**101**:440-445. DOI: 10.1021/jp961744v

[106] Jayanthi SS, Ramamurthy P. Excited singlet state reaction of phenosafranine with electron donors Role of the heavy-atom effect in triplet induction. Journal of the Chemical Society, Faraday Transactions. 1997;**94**:1675-1679. DOI: 10.1039/A800455B

[107] Chaudhuri R, Sengupta PK, Mukherjee KR. Luminescence behaviour of phenosafranin in reverse micelles of AOT in n-heptane. Journal of Photochemistry and Photobiology, A: Chemistry. 1997;**108**:261-265. DOI: 10.1016/S1010-6030(97)00091-9

[108] Broglia MF, Bertolotti SG, Previtali CM. Proton and electron transfer in the excited state quenching of phenosafranine by aliphatic amines. Photochemistry and Photobiology. 2007;**83**:535-541. DOI: 10.1562/2006-07-31-ra-989

[109] Bertolotti SG, Previtali CM. The excited states quenching of safranine T by p-benzoquinones in polar solvents. Journal of Photochemistry and Photobiology, A: Chemistry. 1997;**103**:115-119. DOI: 10.1016/s1010-6030(97)85300-2

[110] Albiter E, Alfaro S, Valenzuela MA. Photosensitized oxidation of 9, 10-dimethylanthracene with singlet oxygen by using a safranin O/ silica composite under visible light. Photochemical & Photobiological Sciences. 2015;**14**:597-602. DOI: 10.1039/ c4pp00261j